BIRDS of the
HIMALAYAS

Bikram Gr

B L O O M S B U R Y
LONDON · OXFORD · NEW YORK · NEW DELHI · SYDNEY

POCKET PHOTO GUIDE

Bloomsbury Natural History
An imprint of Bloomsbury Publishing Plc

50 Bedford Square
London
WC1B 3DP
UK

1385 Broadway
New York
NY 10018
USA

www.bloomsbury.com

BLOOMSBURY and the Diana logo are trademarks of
Bloomsbury Publishing Plc

First published by New Holland UK Ltd, 1998 as *A Photographic Guide to
Birds of The Himalayas*
This edition first published by Bloomsbury, 2016

British Library Cataloguing-in-Publication Data
A catalogue record for this book is available from the British Library.

Library of Congress Cataloguing-in-Publication data has been applied for.

ISBN: PB: 978-1-4729-3826-8
ePDF: 978-1-4729-3824-4
ePub: 978-1-4729-3827-5

2 4 6 8 10 9 7 5 3 1

Designed and typeset in UK by Susan McIntyre
Printed in China

To find out more about our authors and books visit www.bloomsbury.com.
Here you will find extracts, author interviews, details of forthcoming events
and the option to sign up for our newsletters.

CONTENTS

INTRODUCTION

The Himalayas, the world's highest mountain range, form an important Endemic Bird Area. This book deals with birds to be found in the Himalayan range, and aims to demonstrate how species-rich this region is. It covers an area extending from the extreme eastern Hindu Kush mountains of north Pakistan, on to Jammu and Kashmir, the Ladakh area, Himachal Pradesh, Kumaon and Garhwal in Uttar Pradesh, and the terai foothills and Nepal in the central Himalayas, to Sikkim, Bhutan and north Arunachal Pradesh in the eastern Himalayas.

This book covers 252 species of Himalayan birds, both resident and migratory, occurring from the foothills up to the higher elevations. Warm-blooded mammals residing at lofty altitudes possess a larger lung capacity than other species of the plains, enabling them to cope with the paucity of oxygen. Typical high-altitude birds also exhibit an increased lung capacity, and they tend to seek their food mostly on the ground and to fly much less. To protect themselves from harsh weather conditions such as violent gales and blizzards, they very often roost and nest on sheltered ledges or in holes in cliffs.

Owing to their high dependence on specialized food, birds respond adversely to climatic and environmental changes. Since food availability alters according to season and climate, and has recently been affected also by environmental degradation and destruction of natural habitats, birds are forced to migrate to find an alternative food supply.

Many birds inhabiting the higher zones of the Himalayas are constrained to move in winter to the lower valleys, where food is available and temperatures are less harsh. Birds such as pigeons, finches and some raptors fall mainly into this category. Insectivorous birds, such as the flycatchers and warblers, migrate to the lowlands.

Many species included in this book are resident in the Himalayas and undertake only altitudinal movements, dependent upon the season. Others migrate from the plains to the Himalayas in summer in order to breed. There is also increasing evidence suggesting that some birds breeding in the Palearctic region migrate to winter in the lower Himalayas, whereas others can be found only during this particular time in the mountains.

The Himalayan range was once broadly covered by temperate forests, including mixed broadleaves, oak, rhododendron and dry coniferous forests of pines and firs. Such problems as deforestation, drainage of wetlands, and changes in farming techniques in these sensitive hill and mountain areas are the main causes of the rapid decline in appropriate natural bird habitats. As a result, the Western Tragopan, the Cheer Pheasant and the Kashmir Flycatcher are three species that have been so severely threatened by habitat loss and environmental deterioration that they are thought to be at risk of extinction. Moreover, species such as pheasants and partridges are additionally threatened by overhunting and poaching.

It is a fact that, in the last hundred years, an increasing number of bird species have become extinct or highly endangered. The rediscovery of Jerdon's Courser in 1986, however, confirmed Salim Ali's belief that the bird may have survived the enormous pressures placed on India's wilderness areas. Salim Ali also hoped that the Himalayan Quail, which has not been seen since 1876, might have

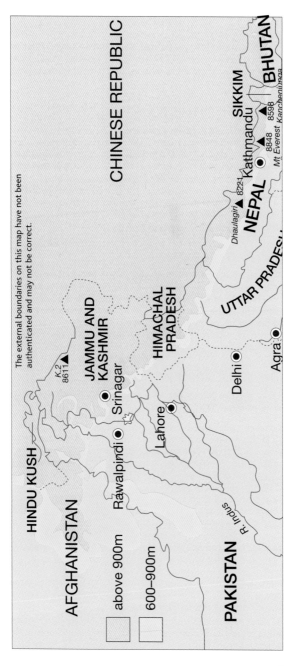

The external boundaries on this map have not been authenticated and may not be correct.

CHINESE REPUBLIC

BHUTAN

SIKKIM

Kathmandu

NEPAL

Dhaulagiri ▲ 8221

▲ 8848
Mt Everest

▲ 8598
Kanchenjunga

UTTAR PRADESH

HIMACHAL
PRADESH

JAMMU AND
KASHMIR

K.2
8611 ▲

● Srinagar

● Delhi

● Agra

Lahore

HINDU KUSH

AFGHANISTAN

● Rawalpindi

R. Indus

PAKISTAN

above 900m

600–900m

gone unnoticed in the western Himalayas. Today, there is much greater awareness of environmental concerns, and it is possible that the thick mountain forests of Bhutan and the eastern Himalayas may reveal new species or subspecies to the ornithologist at the beginning of the 21st century.

HOW TO USE THIS BOOK

In the early 1990s there was considerable confusion over bird nomenclature in this region. The confusion was resolved to a great extent by the publication in 1996 of *An Annotated Checklist of the Birds of the Oriental Region* by Tim Inskipp, Nigel Lindsey and William Duckworth (published by the Oriental Bird Club). However, as this book has not been updated the names in this edition follow Craig Robson's *A Field Guide to the Birds of South-East Asia* (2011, Bloomsbury), which draws from, among other works, *The Howard and Moore Complete Checklist of the Birds of the World* and *Handbook of the Birds of the World*.

THE PHOTOGRAPHS

Each species included in this book is accompanied by at least one colour photograph. For many species the plumage of the male and female are identical, and identification from the photograph should therefore present no problem. In some other species, however, males and females differ in plumage, or the plumage may even change between the seasons of the year. In such cases we have depicted the male, but occasionally a photograph of the female has also been provided.

THE SPECIES DESCRIPTIONS

The descriptions provide detailed information on each species included in the guide, as follows.

Common name As stated above the names in this edition follow *A Field Guide to the Birds of South-East Asia*, *The Howard and Moore Complete Checklist of the Birds of the World* and *Handbook of the Birds of the World*. At present there is still some confusion in this matter, different authors using different sets of names.

Scientific name Each species has a Latin-based scientific name, recognized the world over. There have been some changes in recent times, but here we have followed the same references.

Length After the scientific name, the approximate length in centimetres, from bill tip to tail end, is given.

Main text In the text, reference is made to features and plumage details appropriate to each species. Where space allows, further information is provided on the following aspects.

1. *Flight* because birds are often seen in flight, this important feature, where relevant, is described with reference to the size and shape of the wings, colours and patterns.

2. *Voice* voice can be extremely important in identification, and among the passerines – the so-called 'perching birds' or 'songbirds' – diagnostic songs are often delivered at the start of the breeding season; a much wider variety of birds, including the passerines, have distinctive calls, and experienced birdwatchers are able to identify a large proportion of common birds by sound alone.

3. *Behaviour* each bird species has characteristic behaviour patterns governing aspects of day-to-day life, such as feeding, courtship, response to predators, and whether the bird is solitary or gregarious.

4. *Habitat* generally speaking, birds are faithful to a particular habitat which caters to all their daily needs of food, shelter and protection.

5. *Abundance* most of the birds in this guide are common at least somewhere within the Himalayan range.

Coloured tabs provide an at-a-glance reference relating to the species' family groups. *See* key on following page.

BIRD STRUCTURE AND APPEARANCE

The different species of bird vary greatly in terms of both size and shape. Despite the apparent dissimilarity between, say, a tit and a White Stork, both these species, and all others, have many features in common. The basic plan of the skeleton will be essentially the same; even the 'layout' of the plumage will be similar, the feathers growing in tracts which all birds have in common.

A bird's feathers serve a variety of functions. The most obvious of these have to do with flight. The primaries and secondaries of the wings provide much of the locomotion needed for flight, while the coverts on the wings and the contour feathers on the body aid streamlining and an aerodynamic shape.

Feathers also serve to provide the insulation needed to maintain the bird's body temperature, and in this they are assisted by the underlying down feathers on the body. Lastly, the feathers are often coloured, the patterns produced providing camouflage in some species and a colourful display in others.

The illustration below shows a stylized drawing of a typical bunting. The feathered areas of the head can be clearly seen, as can the feather

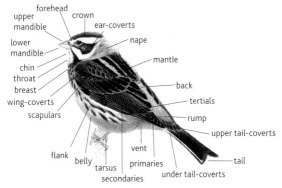

forehead
upper mandible
crown
ear-coverts
lower mandible
nape
chin
throat
mantle
breast
back
wing-coverts
tertials
scapulars
rump
upper tail-coverts
flank
belly
vent
tarsus
primaries
tail
under tail-coverts
secondaries

KEY TO COLOURED TABS

Fowl-like birds

Ducks

Woodpeckers & barbets

Hornbills

Kingfishers & bee-eaters

Cuckoos

Parrots & parakeets

Owls

Pigeons & doves

Rails, crakes & sandgrouse

Waders, gulls & terns

Birds of prey

Bitterns

Leafbirds & allies

Jays, magpies & crows

Orioles & cuckooshrikes

Drongos

Flycatcher-like birds

Dippers, thrushes & relatives

Starlings & mynas

Nuthatches & creepers

Tits

Swallows & martins

Bulbuls

White-eyes

Warblers

Babblers & relatives

Larks

Sunbirds

Sparrows, wagtails, accentors, finches & buntings

tracts of the wings when held at rest. In this posture, inevitably, some of the wing feathers are partially concealed. The relative lengths of the different feathers and the presence or absence of such features such wingbars on particular tracts of feathers can be important in identification.

Study the illustration and try to memorize the names and locations of the most important parts of the bird's body. Feather tracts are often represented by discrete blocks of colour, which can vary from species to species or according to seasonal plumage. A knowledge of this feathering will enable you to make direct comparisons between species and to refer more easily to the descriptions in the text.

IDENTIFYING BIRDS

Identifying birds can sometimes be a very frustrating experience, especially for the beginner. When seen by seasoned observers and under ideal conditions, most of the birds in this book should be readily identifiable, but birdwatchers new to the pastime may have difficulty under some circumstances. Reassuringly, your ability will without doubt improve over the years, and comfort may be taken from by the knowledge that, however experienced the birdwatcher, there will always be some sightings that defy identification.

Your birdwatching skills will improve with time and experience. Here are some of the important aspects of birdlife and birdwatching on which to concentrate.

1. *Size* Since measuring the length of a bird in the field is clearly not an option in most cases, the birdwatcher must learn to gauge size. This is not always so easy as it sounds, as optical aids – binoculars and tele-scopes – can often distort the apparent size. Try to compare the bird in question with a nearby individual of a known species or with an object of known size, remembering to make allowances for foreshortening in perspective if the two are not in the same plane of focus.

2. *Shape* Birds of particular family groups often share a distinctive shape. For example, herons and egrets are, by and large, long-legged and long-necked birds, while plovers have a short, stubby bill, a dumpy body and longish legs. The general shape of the body and the relative sizes of the head, bill and legs are obviously important. The bird's stance and posture should also be noted. Some birds perch upright, while others adopt a more horizontal posture; non-perching birds also have a range of postures and stances, and it should not be forgotten that these can vary according to the behaviour of the bird.

3. *Colour* When identifying a new bird, make careful observations of the patterns and colours of the plumage. Lighting can be crucial in the observation of colour, so take this into account. Also bear in mind that plumage can vary according to the time of the year, age and the sex of the individual, so you need to analyse whether you are looking at a bird in breeding plumage or a non-breeding bird in winter plumage. Juveniles, which often have cleaner-looking feathering than adults in the autumn, are invariably different in appearance from their parents.

4. *Behaviour* The behaviour of a bird can give vital clues to its identification, although a degree of patient observation may be needed to discern the more important patterns. Watch, for example, the way in which the bird feeds, whether on the ground or among foliage, or its flight pattern and its response to others of the same or a different species.

5. *Habitat* Each species is perfectly adapted to feed in a characteristic way in a particular habitat, and only under exceptional circumstances will it be found elsewhere. There are, of course, species which are exceptions to the rule. With most bird species, a knowledge of their habitat preferences can greatly improve the chances of being able to find them.

6. *Voice* When it comes to a knowledge of bird songs and calls, there is no substitute for field experience. Most birdwatchers will in time learn about all the more common and widespread species. This can be very satisfying and, at the very least, provides you with a basis for comparison when dealing with unfamiliar calls.

7. *Jizz* The 'jizz' of a bird is a combination of all the features and aspects discussed above, and others, seen through the eyes of an experienced birdwatcher. Given years of observation, some people can identify birds even in poor light and at a considerable range. Clues might be the flight pattern of a speck on the horizon, or the feeding pattern of a distant wader far out on the mudflats. At least part of the skill, however, derives from having the background knowledge to make a 'best guess' at the most likely species in a particular habitat at a certain time of year.

GOING BIRDWATCHING

Having familiarized yourself with the layout of the guide, it is now time to go birdwatching. Although, at the simplest level, all you need is a pair of eyes, some expenditure on equipment can greatly increase the pleasure derived from the pursuit. In addition, the development of fieldcraft skills and a knowledge of when, where and how to watch birds can enhance the experience.

Equipment A pair of binoculars is part of the uniform of every birdwatcher (along with waterproof boots and clothing). It goes without saying that, the more you spend, the better the quality of the binoculars. There are numerous models on the market with price tags ranging from under £100 to nearly £2,000. You do not necessarily have to spend a fortune to acquire a good pair, so visit a reputable dealer and try out a few before you make a purchase.

All binoculars bear a set of numbers which indicate their likely value to the birdwatcher. A typical pair may have the numbers 8×40 on the casing. The first number is the degree of magnification of the lenses, and the second is a measure of their light-gathering capacity and hence the brightness of the image. Do not be tempted into purchasing anything with a magnification greater than 10 or less than 6. The second number should not be less than 30, or the image will be too dark; nor perhaps more than 50, since the binoculars may then be too heavy to hold still.

The best advice would perhaps be to purchase the binoculars with which you feel most comfortable. The same goes for telescopes, which nowadays also come in all shapes and sizes, and you also need a tripod on which to stand a telescope for prolonged observation. The dedicated birdwatcher soon becomes laden with all manner of paraphernalia!

When and where to go birdwatching Time of day, time of year and weather all have an important bearing on the variety and numbers of birds in a given habitat. Before planning an outing that involves any travel, it is often wise to sit and consider your options.

Habitats offer birdwatching opportunities that vary throughout the year. Woodlands, for instance, are often at their best in the spring, when territorial birds are in full song and the leaves are not fully open; although birdwatching is good throughout the morning, arrive as soon after dawn as you can for the best chorus. Woodlands in summer are often rather uninspiring for the birdwatcher; but, perhaps surprisingly, winter has much to recommend it, with mixed flocks of small birds roaming the trees and the lack of leaves making observation easy.

While many common species are spread over large areas, others are limited not just to a region but also to a habitat. Some birds of the conifer forests of the hills, for example, will be found nowhere else but there. City gardens are homes for many species, including tailorbirds, sunbirds, white-eyes, babblers and mynas. Shallow lagoons, inland jheels or shallow lakes and rivers are rich habitats for waterbirds such as pelicans, storks, cranes and egrets.

Most birdwatchers are lured to the hills, where many of the birds are distinctive. Typical of these are the tits, accentors and finches. Again, different areas of hills are best at different times of year. Higher reaches are best in spring and summer, while the lower foothills come into their own in autumn and winter.

Acknowledgements

The author would like to thank the official team, led by Aruna Ghose and consisting of Rajinder Bist, Madhulita Mohapatra, Ajay Verma and Pramit Chanda; Ria Patel for her help in researching this book and for being the only young person he knows who has shown some inclination to take it on from here; Tim and Carol Inskipp, Nigel Redman, Krys Kazmierchek, Sudhir Vyas, Per Undeland and Otto Pfister for insisting that he take a more scientific approach to his birdwatching; Rattan Singh and Yogi and Bittu Sahgal for being constant companions in the field; and Miel, Mimi, Myrti, Tara, Aamana, Nikhat and Samiha for being his brood.

Otto Pfister would like to thank Sanjeeva Pandey, Dhanraj Malik, Tashi, Vinod and Ramesh for their company, Bikram Grewal, Aruna Ghose and Ria Patel for their input into the book, and Per Undeland and Tim and Carol Inskipp for help with photographs. All the photographs in this book were taken by Otto Pfister unless otherwise captioned.

SNOW PARTRIDGE *Lerwa lerwa* 38cm

Joanna Van Gruisen

A medium-sized partridge, barred black and white above, deep chestnut below, with broad white streaks on abdomen and flanks; chestnut undertail, streaked black and tipped white. Social; coveys of 5–20 may be seen on alpine pastures and among bushes. Rather tame when not harassed. Has a loud call, rather like that of the Grey Francolin (*Francolinus pondicerianus*). Feeds on seeds, vegetable shoots, lichens, moss. An endangered species; frequents alpine meadows, scrubby hillsides, rhododendron and fern undergrowth. Resident in the Himalayas, between 2500m and 5000m.

TIBETAN SNOWCOCK *Tetraogallus tibetanus* 50cm

A big, dumpy partridge with strong, shortish red legs, its colours merging with surroundings. Crown and upperparts dark grey, with white-streaked grey wing-coverts, and rufous uppertail-coverts and central tail; forehead, ear-coverts and throat white, with dark grey bands on neck sides and breast; otherwise white below, streaked with black. Male heavier than female. Small groups give constant soft clucking calls; shrill cackles when alarmed. Male very vocal when breeding. Digs vigorously with strong yellow-red beak for tubers, roots and shoots; swallows grit. Resident of high-altitude alpine meadows and boulder-strewn grassy hillsides in central-western Himalayas (Ladakh; Rupshu, northern Nepal), to 5800m in summer; descends in winter to 3000–4000m, lower when snowfall heavy.

HIMALAYAN SNOWCOCK *Tetraogallus himalayensis* 65cm

Joanna Van Gruisen; Fotomedia

A large partridge, well camouflaged in its preferred haunts. Grey, white, black and chestnut, streaked and blotched; white throat is bordered by broken chestnut collar; dark grey below breast, with white vent. Trailing edges of wings appear transparent in overhead flight, especially against a bright sky. Noisy, the cock uttering loud whistle of several notes. Small parties of 3–10 rummage on hillsides, scratching and digging furiously for grass shoots, tubers and bulbous roots; sometimes swallows grit. Tends to fly downhill or to run up when alarmed. Found on alpine meadows and rocky terrain above treeline. Resident from Kashmir to Kumaon and western Nepal, at 3800–5500m in summer; descends to about 2200m in winter.

CHUKAR PARTRIDGE *Alectoris chukar* 35cm

A medium-sized, red-legged partridge with uniform grey-pinkish-olive upperparts and prominent rib-like bands on flanks. A black band across forehead, through eyes and across breast, enclosing bright white cheeks and throat. Female slightly smaller, lacks spur on tarsus. The vernacular name is onomatopoeic, reflecting the call. Occurs in family groups, occasionally in larger numbers, scattering when flushed. Fast on the wing, but tends to run before taking flight. Found on barren or open mountainsides. Resident throughout the western Himalayas, between 1200m and 5000m.

TIBETAN PARTRIDGE *Perdix hodgsoniae* 30cm

A typical stocky partridge with short, green-brown legs. Rufous from nape to shoulder; face white, with distinct black cheeks; back tawny brown-grey, partly rufous or dark-barred, tail with rufous at sides; whitish-grey below, barred black from lower throat over breast, belly and flanks, which also have rufous streaks. Strong pale green bill. Voice as other partridges; calls mainly in morning and late afternoon. Feeds in pairs, or small groups in winter, mainly on seeds, shoots and insects. Just crosses Tibetan border into northernmost west-central Nepal, eastern Ladakh and Sikkim. Locally not uncommon, in summer between 3600m and 5600m; in winter generally below 4000m, rarely as low as 2800m.

COMMON QUAIL *Coturnix coturnix* 20cm

Male pale brown above, boldly streaked and marked, with blackish chin and stripe down throat centre and narrow stripe curving to ear-coverts; breast rufous-buff with white shaft streaks; abdomen is whitish. Female has buff throat and breast heavily streaked with black. Male's liquid whistling *wet-me-lips* is a common and familiar call. Found in pairs and small parties on ground, often huge numbers at favoured feeding sites; it eats seeds, grain and insects. Well-known resident, in cultivation, crops and grasslands; commoner in winter, when numbers augmented by migrants. Breeds in Kashmir to about 2500m, and in parts of north and north-east India.

Göran Ekström

HILL PARTRIDGE *Arborophila torqueola* 28cm

Peter Morris

Male is chestnut from forehead to nape, with black eyebrows and lores and rufous-buff ear-coverts. Olive-brown above, streaked and mottled black and chestnut; black throat and neck streaked white; prominent white band between foreneck and breast. Female has buff-rufous throat and neck sides, with black streaks on sides of head and neck. Has mournful whistle, *po-eer po-eer*, the second syllable slightly longer. Small parties feed on forest floor, on seeds, insects, berries; roosts in trees at night, several huddled together. Found in dense jungle undergrowth in the Himalayas, east of Garhwal through Nepal, Sikkim and Bhutan to extreme eastern Assam, at 1000–4000m.

BLOOD PHEASANT *Ithaginis cruentus* 46cm

A gregarious bird, the male bright red and black about the face, with yellow mop-like crest. Streaky greyish above; apple-grey below, thickly streaked yellow; crimson on upper breast, wings and tail. Female rich rufous-brown, finely marked, with scarlet around eyes. Has long-drawn squeal. A strong runner, but rarely flies; can be tame. Feeds in clearings, on fern and pine shoots, lichens, moss. Resident in steep hill-forests of pine and dwarf rhododendron, dense ringal bamboo and juniper scrub; prefers snow-covered areas. Occurs at high levels in the Himalayas, east of central Nepal, at 3200–4300m, moving up and down seasonally with the snowline.

WESTERN TRAGOPAN *Tragopan melanocephalus* 71cm

A little-known bird of dense forest undergrowth. Male is crimson and black, profusely spotted with white, with red-tipped crest and red face patch, deep blue featherless throat, and reddish on upper breast. Female grey-brown, with rufous head and neck, and black and white streaks and spots on upperbody. An unmistakable goat-like *waa-waa-waa* call. Solitary or in pairs. Shy and elusive, but occasionally emerges to feed in open, with other pheasants, around melting snow patches; eats fresh leaves, bamboo shoots, seeds, berries, also insects. Inhabits forests in the west Himalayas.

SATYR TRAGOPAN *Tragopan satyra* 68cm

The handsome male is rich orange-red, with olive-brown on back and rump, bestrewn all over with black-bordered white spots, with a black crest streaked with crimson on each side. Female duller, a rufous-brown or ochreous-brown, barred and blotched with black and buff. Has a loud *wak* call, repeated several times, or a loud *kya-kya, kya*, rather like bleating of a goat. Resident of oak, deodar and rhododendron forests in the Himalayas, between 2400m and 4250m; descends to lower valleys during severe winters.

16

KOKLASS PHEASANT *Pucrasia macrolopha* 60cm

Male is silvery-grey above, streaked black, and deep chestnut below, with metallic green head and horns, a long brown occipital crest, and a pointed chestnut-brown tail. Female (55cm) is black and brown, mottled and streaked, with buffy-chestnut crown and short crest. Best located by its loud *khok-kok-kok-kokha* call; vocal around dawn and dusk, but intermittently through day when breeding (April-June). Pairs or small parties keep to steep slopes; difficult to flush, and then flies up at great speed before plunging down. Emerges in clearings in early morning. Eats tubers, shoots, leaves, seeds, insects. Found on steep forested hills and ravines from southern Kashmir to central-east Nepal, at 1500–4000m, coming lower in winter.

Sanjeeva Pandey

HIMALAYAN MONAL *Lophophorus impejanus* 72cm

The 'pheasant of nine colours', the male being black below, with metallic green head and crest becoming metallic purple from mantle to and around the rufous neck, and with white rump, metallic green uppertail-coverts, rufous-chestnut tail and black wing quills. Female has shorter crest than male, and is brown, mottled and patterned all over, with white throat; white in tail prominent in flight. A whistling call in flight. When flushed, flies into trees and freezes. Occurs in high-altitude forest edge and open forest between 1800m and 5000m, moving up and down within this range depending on season.

KALIJ PHEASANT *Lophura leucomelanos* 70cm

This typical pheasant is a close relative of the Silver Pheasant (*Lophura nycthemera*). Male glossy black above and brownish-grey below, with rump feathers white-edged, also some white below, and with bare red patch around eye; crest long and black (in the nominate race) and tail long and drooping. Female reddish-brown with paler scalloping, and with stiff tail. Various chuckles and clucks uttered by both sexes. Wanders out to open ground to feed on seed and grain, but never too far from cover. Found in thick undergrowth of steep mountainside forests, often near water, throughout the Himalayan foothills to 3500m, coming lower in winter.

CHEER PHEASANT *Catreus wallichii* 90cm

Male is pale buff above, closely barred black, with a dark brown head with long crest and bright red naked orbital patch, and buffy-white below, with black belly. Female (70cm) more chestnut below, and with paler orbital patch. Has very loud, distinct call, *chir-pir chir-pir*, also some cackling calls; noisy before dawn and at dusk. Very shy and skulking; hurtles downhill when flushed. Feeds on seeds, roots, tubers and insects. Inhabits grass-covered steep, rocky hillsides with scattered trees. Resident in the Himalayas, from north-west Pakistan to central Nepal, at 1200–3500m.

Sanjeeva Pandey

COMMON MERGANSER *Mergus merganser* 66cm

A fish-eating diving duck, having a hooked bill with serrated edges. Breeding male is white with black head, back and primaries. Other plumages grey-brown above, white below, with rufous head and neck sharply demarcated from body, and bold white speculum; female has short crest, and non-breeding male has a white line from lores to bill. Ungainly on land, but expert swimmer; often hunts fish cooperatively in cormorant fashion. Tends to fly low over water. Frequents large waterbodies, both still and flowing. Rare within our limits, though common in Ladakh and winters in the east Himalayan foothills.

EURASIAN WRYNECK *Jynx torquilla* 19cm

A small, inconspicuous bird, overall greyish-brown with a dark patch on mantle extending to central nape/crown, and barred tail, throat and upper breast. Has a distinctive call, rather like that of a small falcon. Seen either singly or in pairs (during breeding season). Feeds on the ground; opens anthills with its short, pointed beak in search of its preferred diet of ants, mainly larvae and pupae. Flies in shallow undulations. Occurs in open forests, clearings and woodland with low undergrowth. Winters in north India from early September to end April.

BROWN-CAPPED PYGMY WOODPECKER
Dendrocopos nanus 12cm

A tiny, active woodpecker, barred brown and white above, with paler crown, and prominent white band extending from just above eye to neck, and dark-streaked pale brown-white below. Male has short scarlet streak on side of rear crown. Has a faint but shrill squeak. Mostly in pairs or with mixed bird parties in forest. Seen more on slim trees, branches and twigs, close to ground (also high in canopy). It obtains small insects and grubs from crevices and under bark; also eats berries. Resident of light forests, bamboo and cultivation from Rawalpindi district in Pakistan, east through Rajasthan, Uttar Pradesh and the Nepal terai and foothills.

GREY-CAPPED PYGMY WOODPECKER
Dendrocopos canicapillus 14cm

A small woodpecker with boldly white-barred black upperparts, ash-grey forehead and crown, and white neck sides extending into broad whitish supercilia. Crest scarlet surrounded by black on male, but all black on female. Partial to investigating thin twigs, branches and stems, either high in canopy or low down near ground, which are rarely used by larger woodpeckers. A frequent member of mixed-species flocks of insectivorous birds. Otherwise, a typical woodpecker in all respects. Resident in Himalayan open and secondary forests.

RUFOUS-BELLIED WOODPECKER
Hypopicos hyperythrus 20cm

Rupin Dang

A quiet woodpecker, the male heavily red on head, this colour extending to hindneck and neck sides, with underparts a deep chestnut. Female has forehead to hindneck black with prominent white spots. Likes to drum on dead stumps. Observed singly or in separated pairs, often as part of a roving band of insectivorous birds, feeding on ants, beetle grubs and other insects usually fairly high up on the trunks of large trees. Resident in high-altitude forests, mainly from Nepal eastwards to the eastern Himalayas.

HIMALAYAN WOODPECKER *Dendrocopos himalayensis* 25cm

A myna-sized woodpecker, both sexes having a black back with a prominent elongated white patch on the shoulders and white spots and bars on the flight feathers, and white cheek and ear-covert patch bordered with black. Crown and small crest crimson on male, black on female. Yellow-brown to dirty white below, this colour extending up as a band at base of neck; vent crimson. Voice and habits very like those of the familiar Great Spotted Woodpecker (*Dendrocopos major*) of Europe and other parts of Asia. Resident of the cold, high Himalayan forests.

RUFOUS WOODPECKER *Micropternus brachyurus* 25cm

Krupakar-Senani: Sanctuary Photo Library

An inconspicuous woodpecker, chestnut with fine black bars on upperbody, including wings and tail, and paler-edged throat feathers. Male has crimson patch under eye. Rather vocal in January–April: loud, high *ke-ke-ke-ke*, like a myna's; drums during breeding period. Pairs, sometimes loose groups of 4–5, mostly seen at ball-shaped nests of tree-ants, clinging to outside, digging for ants and their pupae; plumage, especially head, breast and tail, often smeared with gummy substance. Also eats fruits, and seen taking sap from near base of banana leaves. In mixed forests south of the outer Himalayas, east from Dehra Dun, at up to 1500m, but commoner at lower altitudes.

GREATER YELLOWNAPE *Chrysophlegma flavinucha* 33cm

A largish bird, mostly dark green, usually paler or greyish below, with yellow crest and hindneck. Male has prominent yellow throat, replaced by chestnut on the female. Flight feathers barred rufous and black. Active, noisy, maintaining constant vocal contact; various *ke-ep* notes, also *chup-chup*. Shy, restless, found in pairs or loose family parties; it may associate with drongo, babbler and bulbul species. Long, slightly chisel-tipped bill suited to picking ants, termites and large insect larvae; also takes berries and seeds. Found in open evergreen and deciduous forests on the lower Himalayan slopes.

STREAK-THROATED WOODPECKER
Picus xanthopygaeus 30cm

A little bigger than a myna, this woodpecker is grass-green above with a yellow rump, and greenish below with grey chin and throat. Underparts heavily scalloped, and remiges banded and spotted. Male has red crown, and female black; both have a white malar stripe and a black-bordered white supercilium. As well as the typical woodpecker habit of working on tree trunks and branches, chiselling for wood-boring insects, this species comes to the ground to feed on ants and termites. Resident of open forests in the Himalayas, to about 1500m, occasionally 1700m.

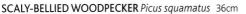

SCALY-BELLIED WOODPECKER *Picus squamatus* 36cm

A typical woodpecker. Green above with yellow rump, and heavy scaly pattern from lower breast downwards; wings and tail barred. Male has forehead to nape red, bordered with black, female has top of head all black; both have white supercilium and black moustache, and long, straight, chisel-tipped bill. Characteristic call: ringing *klee-gu* or similar, also nasal *peer*. Singly or in pairs, moving up trunks and branches in jerky spurts. Takes ants, termites, larvae of wood-boring insects, and berries in winter. In forests, both deciduous and evergreen, along the Himalayas from extreme west to central Nepal, at 1000–3700m.

GREY-HEADED WOODPECKER *Picus canus* 33cm

Male is darkish green above, with crimson forehead, black hindcrown and nape, black malar stripe, yellow rump, white-barred dark wings and blackish tail; unmarked dull greyish-olive underbody diagnostic. Female has forehead to nape black, with no crimson. Usual call a high-pitched *keek-keek...* of four or five notes; drums often between March and early June. Singly or in pairs, moves along trunks and larger branches, seeking insects under bark; descends to ground, hopping awkwardly, and digs into termite mounds. Food includes ants, wood-boring beetles and their larvae, nectar, fruits. Inhabits the Himalayan forests, from lower foothills to about 2700m.

GREAT BARBET *Megalaima virens* 33cm

Our largest barbet, this bird has a blackish head, brownish-olive back and shoulders, the upper back streaked with yellowish and the wings and tail with green and blue, and a prominent yellow beak. Below, the upper breast is dark olive-brown and the belly striped yellow and brown with a blue median band; vent conspicuously scarlet, undertail blackish. A far-carrying loud, plaintive *mee-ou* is repeated continuously in the breeding season. Keeps to canopy of tall trees, unless coming lower to feed on fruit. Flight undulating, rather like a woodpecker's. Inhabitant of temperate moist forests of the Himalayas.

LINEATED BARBET *Megalaima lineata* 28cm

A myna-sized, arboreal bird with grass-green plumage, broadly pale-streaked above and below. Has a naked yellow patch around eye, not extending to the base of its dumpy, stout bill. Very noisy in hot weather; call similar to that of the Brown-headed Barbet (*Megalaima zeylanica*), though softer and more mellow. Can be seen singly or in loose feeding parties on fruiting trees. Feeds on all kinds of fruit, nectar, insects, termites, larvae and grubs. Shares similarities with other barbets, particularly the Brown-headed Barbet. Resident in the submontane Himalayas, from the edge of the terai through the foothills up to 1000m.

BLUE-THROATED BARBET *Megalaima asiatica* 24cm

A myna-sized bird, grass-green overall, with a pale blue face and a crimson crown. The crimson hindneck is enclosed within a black band, this in turn bordered with yellow at front; a large crimson spot on each side of neck; remiges blackish, underparts yellower. Shares its dumpy appearance and general habits with other barbets: arboreal and frugivorous, with undulating flight; also typical monotonous call, three-syllabled in this species, *kutooruk-kutooruk*, uttered for long periods. Inhabits light forest and well-wooded country of the lower Himalayas, from west Pakistan, Punjab and Kashmir eastwards.

Joanna Van Gruisen

ORIENTAL PIED HORNBILL *Anthracoceros albirostris* 90cm

Nigel Redman

A largish hornbill. Black above, with white face patch, wingtips (seen in flight) and tips to outer tail feathers; white below, with black throat and breast. Beak yellow, with front part of the large casque partly black. Female slightly smaller. Cackles loudly and screams; also a rapid *pak-pak-pak*. In small parties, occasionally several dozen together on favourite fruiting trees; associates with other birds. Arboreal, but often feeds on ground; food includes fruits, lizards, snakes, young birds, insects. Found in open forests, plains, orchards and groves of the lower Himalayan foothills, from Punjab and Kumaon to extreme north-east and central India.

GREAT HORNBILL *Buceros bicornis* 130cm

Joanna Van Gruisen

The largest of our hornbills, this vulture-sized, endangered species has a huge yellow (partly black) bill and casque, the latter forked at front. Tail white, with broad black band more towards end; face and upperparts black, with white neck and wing patches. Female slightly smaller, with posterior end of casque red. Iris of male blood-red, of female white. Keeps in small groups, flying from tree to tree in almost 'follow-my-leader' fashion, maintaining rigid punctuality in movements around home range. Restricted to the wet and moist forests of the Himalayas and the north-east, at up to 2000m.

BLACK-BACKED KINGFISHER *Ceyx erithaca* 13cm

A tiny kingfisher, and perhaps the region's most beautiful. Bright coral-red legs and feet; the head and neck are orange-rufous with violet-blue and white patch, and lilac mantle with blue streak down back; dark wings; underparts are orange-yellow with white chin and throat. Has a sharp, squeaky call, *chicheee*. Catches small fish, crustaceans and the like from shaded forest streams; also picks insects from forest floor. Other general habits as those of the Common Kingfisher (*Alcedo atthis*), including darting off when approached. Found in deep evergreen and moist deciduous forests, from Nepal eastwards.

Joanna Van Gruisen; Fotomedia

STORK-BILLED KINGFISHER *Pelargopsis capensis* 38cm

A big kingfisher. The greyish-brown cap, large, thick blood-red bill, yellowish collar and pale greenish-blue upperparts, combined with white and brownish-yellow underparts, distinguish it from other kingfishers. Found singly or in separated pairs. Has loud, noisy *ke-ke-ke-ke-ke* call, uttered from perch. Rather secretive, sitting on concealed perch in foliage overhanging water, darting down to capture any animal of manageable size; feeds on fish, frogs and small birds. Frequents calm waters in wooded habitats in the lower Himalayas.

CRESTED KINGFISHER *Ceryle lugubris* 40cm

Joanna Van Gruisen

A huge black and white kingfisher, the largest found in the region. Wings and tail boldly barred and spotted with white. Head with prominent erect crest; lacks Pied Kingfisher's supercilium; white throat separated from white cheek patch by a thin dark band joining bill with the darker neck sides. White below, with black and rufous-brown breast band, blackish bars on flanks and vent. Female like male, but has armpits and the underwing-coverts are reddish-brown. Usually in pairs; rather uncommon and shy, difficult to approach. Does not hover like its smaller relative. Found on the larger Himalayan streams and confluences.

PIED KINGFISHER *Ceryle rudis* 30cm

A myna-sized black and white kingfisher with a small black crest, prominent white supercilium, and an indistinct white collar on hindneck. Female differs from male in having a single breast band broken in the middle, while the male's is double and complete. Performs incredible hover while fishing: dashes up to some five metres above water surface, suddenly stopping in one position, bill pointing down as if intently taking aim, to plunge headlong into water to take fish; also feeds on tadpoles and water insects. A plains bird always found near water, at up to 2000m in the Himalayas.

BLUE-BEARDED BEE-EATER *Nyctyornis athertoni* 36cm

A pigeon-sized bee-eater, bright green and blue overall, the shades of green varying. Forehead verditer-blue, head and neck tinged with this colour; elongate feathers of foreneck and breast deep blue; rest of underparts washed buffish, vent fully buff; undertail ochraceous-yellow. Bill shorter and thicker than on other bee-eaters. Croaking sound, often followed by softer chuckling call; also a harsh *korrr-korrr*. A shy and rather inactive bird, usually keeping to treetops. Flight undulating. Found in degraded and opened-up evergreen and moist deciduous forest in the lower Himalayas, to 1700m.

CHESTNUT-HEADED BEE-EATER *Merops leschenaulti* 21cm

One of the most beautiful bee-eaters. Head, neck and back chestnut, with wings and tail dark green; yellow throat separated from the green underparts by a dark chestnut band; uppertail-coverts pale blue. Central tail feathers not elongated, the tail appearing squarish. Gregarious like other bee-eaters, roosting communally. Feeds chiefly on winged insects, captured in flight. Found in forested and wooded areas near streams and other watercourses; likely to turn up in any appropriate habitat, especially during the rains, outside its range in the Himalayan submontane tracts.

INDIAN CUCKOO *Cuculus micropterus* 33cm

Rupin Dang

A hawk-like cuckoo, but with much weaker flight. Slaty-brown above, greyer on head, throat and breast; whitish belly with broadly spaced black bars; broad subterminal tail band (characteristic of non-hawk cuckoos of genus *Cuculus*). Female has rufous-brown wash on throat and breast. Solitary, arboreal, not easy to see; call ('one more bottle') the most important clue. Feeds on insects, especially hairy caterpillars. Frequents fairly wooded country in deciduous and evergreen zones. Variously resident, nomadic, or seasonal visitor (during rains and winter), in the lower Himalayas from Kashmir eastwards, at up to 2300m, at times to 2800m.

EURASIAN CUCKOO *Cuculus canorus* 33cm

A slender bird, often sitting still for long periods on branches of thick-foliaged trees. Male's head and chest are ashy-grey, the back darker, becoming blackish-brown towards tail; lower breast to vent heavily barred grey and white. Some females very different, rufous-brown with blackish barring. Strong and fast flier, wing shape and flight pattern like a sparrowhawk. Wary, more easily heard than seen: male gives repeated *kuk-koo* in long unbroken sequences. Takes caterpillars and other insects from ground and leaves. Summer visitor, common in heavily wooded parts in the Himalayas, from Afghan border through Kashmir to Nepal and Sikkim.

DRONGO CUCKOO *Surniculus lugubris* 25cm

Tim Loseby

Resembles a Black Drongo (*Dicrurus macrocercus*), including forked tail. Glossy black, with white-barred undertail and base of outer tail feathers. Young birds dull in colour, speckled white. Voice diagnostic: loud, distinctive call of 5–7 rising notes, a whistling *pee-pee-pee-pee-pee*, ends abruptly, only to begin all over again. Noisy during monsoon or overcast weather, when it disperses widely. Solitary, and strictly arboreal; mostly overlooked and mistaken for drongo, but cuckoo-like flight a giveaway. Feeds on insects and wild fruit. Found in open forests, orchards, cultivation with trees, in the lower Himalayas up to 2000m, but widespread during rains.

VERNAL HANGING-PARROT *Loriculus vernalis* 13cm

An energetic, chiefly arboreal little bird with distinctive short, square tail and bright crimson rump. Otherwise, bright grass-green with small blue throat patch; female lacks blue on throat. Call a faint clucking note. Solitary or in pairs, sometimes in large flocks in flowering and fruiting trees; occasionally seen with other birds in mixed parties. Difficult to locate in canopy, where moves acrobatically around branches or hangs upside-down to feed on nectar, soft fruit pulp and seeds. Sleeps hanging upside-down, like bat. Resident in Himalayan forests and orchards, east of Nepal; moves a great deal locally.

SLATY-HEADED PARAKEET *Psittacula himalayana* 40cm

Vivek Menon

Grass-green, with deep slaty-grey head, black chin and narrow neck ring, blue-green hindneck collar, red shoulder patch, and long, pointed yellow-tipped tail. Female lacks red on shoulders. Has high-pitched but pleasant double-note *tooi-tooi* call, somewhat interrogative in tone, uttered mostly in flight; also a single-note call. Strong flier. Occurs in small flocks; arboreal, but often feeds on standing crops. Prefers fruits, nuts, maize; often causes damage. Found in forests, orchards and hillside cultivation in the Himalayas, to about 2800m; moves around considerably, descending very low in winter.

PLUM-HEADED PARAKEET *Psittacula cyanocephala* 33cm

Relatively smaller and slimmer than other parakeets. Male yellowish-green with plum-red head, black and bluish-green collar and maroon-red shoulder (wing) patch; white tips to central tail feathers distinctive. Female has dull bluish-grey head, yellow collar, and almost non-existent maroon shoulder patch. Call a pleasant tooi-tooi, as well as other notes when perched. Flight direct and swift, the birds rolling from side to side. In pairs or small groups, sometimes joining into much larger gatherings. Feeds on fruit, grain, buds, fleshy petals and nectar. Prefers woodland and forest along the lower Himalayas, through Jammu and Kashmir, Uttar Pradesh and Nepal to Bhutan.

RED-BREASTED PARAKEET *Psittacula alexandri* 33cm

A pigeon-sized, grass-green parakeet with plum-red throat and breast, yellow shoulder patch and long pointed tail. Male has purple-grey head with narrow black band on forehead and upper lores; bill largely red. Female's head is tinged blue-green, breast darker red, and bill largely black. Short, sharp, nasal scream, quickly repeated by several birds together when flock is disturbed. Generally in parties of 6–10. Often descends to ground in harvested fields. Eats wild figs, other wild and cultivated fruit, leaf buds, petals, nectar. Resident of the lower hills from Uttar Pradesh to far eastern Himalayas, to 1500m; local nomadic movements depending on food supply.

ALPINE SWIFT *Tachymarptis melba* 22cm

Göran Ekström

An extremely strong flier, with very long, sickle-shaped, pointed wings. Dark sooty-brown above; white below, with diagnostic broad brown band across breast, and dark undertail-coverts. Loose parties dash about erratically at high speed, feeding on winged insects; may also catch insects disturbed by forest fires. Can be seen high in sky at dusk, many birds wheeling and tumbling, their shrill *chrrrr chee-chee* screams rending the air; utters twittering notes at roost sites. Drinks at ponds and puddles by skimming over the surface. Found in hill country and at cliffsides, above 2300m in the Himalayas.

INDIAN EAGLE-OWL *Bubo bengalensis* 65cm

Slightly larger than a Black Kite *(Milvus migrans)*, this owl is dark brown, streaked and mottled with buff and black. Orange eyes and fully feathered legs distinguish it from similar species. The distinct 'horns' are held rather erect, whereas the Brown Fish Owl holds them flat. Deep, resonant, far-carrying *bu-bo* call, repeated at intervals. Roosts by day in a rock fissure or dense-canopied tree, flying off a good distance when disturbed. Feeds on rodents, fish, reptiles, frogs and medium-sized birds. Inhabitant of vegetated hillocks, ravines, valleys, rocky and eroded areas, at 2000–4000m, from the north-west Himalayas to Himachal Pradesh.

BROWN FISH-OWL *Ketupa zeylonensis* 56cm

A very big owl, dark brown above and strongly streaked below. Further identification aids are a white throat patch and naked legs. Solitary or in pairs. Mostly nocturnal, spending day on leafy branch, rock ledge or in old well; flies slowly but considerable distance when disturbed. Emerges to feed around sunset, advertising its arrival with a characteristic deep, booming *bu-boo* call; snapping calls at nest. Feeds on rodents, fish; also reptiles, frogs and medium-sized birds. Frequents ravines, cliffsides, riversides, scrub and open country. Resident throughout region, from about 1500m in the Himalayas.

HIMALAYAN WOOD-OWL *Strix nivicola* 45cm

A large, brown owl with a greyish facial disc and closely barred and streaked underparts; lacks ear tufts. Nocturnal, perhaps at least partly because of attacks by crows and other diurnal birds during the daytime. Spends the day perched upright and motionless on a branch near the trunk, practically or fully concealed by the foliage. Has a loud *hoo hoo hoo-hoo-hoo* call; also a short, sharp *ki-ick*. Feeds on small rodents, birds and lizards. Resident in oak, pine and fir forests throughout the Himalayas, between 1200m and 4250m altitude.

COLLARED OWLET *Glaucidium brodiei* 16cm

A bulbul-sized bird, barred grey-brown, with bold white supercilium, rufous half-collar on upper mantle and a white throat patch. Utters a pleasant four-note, bell-like whistle. Single birds usually seen perched close to tree trunk, but far less nocturnal than other owls; flies about freely in open sunlight, hunting and calling continuously during middle part of day. Extremely bold and fierce for its size. Feeds on small birds, mice and lizards. Resident in open hill-forests throughout the Himalayas, at up to 3200m altitude.

JUNGLE OWLET *Glaucidium radiatum* 20cm

Darkish brown above, barred rufous and white, with flight feathers barred rufous and black. White moustachial stripe, and white centre of breast and abdomen; remainder of underbody barred dark rufous-brown and white. Lacks ear tufts. Seen singly or in pairs. Noisy, with musical *kuo kak kuo-kak* call. Crepuscular, sometimes also active and noisy by day, but otherwise spends its day in leafy branches; flies a short distance when disturbed. Feeds on insects, small birds, lizards and rodents. A forest bird, partial to teak and bamboo mixed forests, at up to 2000m from Himachal Pradesh east to Bhutan.

LITTLE OWL *Athene noctua* 23cm

This rather silent, small, round-headed owl has coloration perfectly adapted to its hostile, arid environment. Upperparts and upper breast sandy grey-brown, spotted white, with chevron-shaped whitish neck collar; buff below, broadly streaked brown. Golden-yellow eyes and strong yellow bill set in whitish round facial disc, with striking white 'eyebrows'; legs and feet covered with fur-like buff feathers. Call a plaintive *piu*. Hunts even in daytime, perched motionless, suddenly catapulting forward to make precision landing on victim; takes mice, insects, small birds. Found mostly around rocks and boulders on the barren hillsides of the western Himalayas, to 4500m.

BROWN BOOBOOK *Ninox scutulata* 32cm

Chew Yen Fook

A hawk-like owl, slightly larger than a pigeon. Dark grey-brown above, with pale forehead and white patches on shoulders; throat and foreneck brown-streaked fulvous, rest of underparts white with large red-brown patches forming broken bars. Tail barred black, tipped white. Voice very distinctive and diagnostic: *oo-uk*, repeated up to 20 times. Crepuscular and nocturnal. Single birds or close-huddled pairs spend the day on shady, thickly foliaged branches. Feeds on large insects, frogs, lizards, small birds, mice. Forest resident of the outer Himalayas and submontane tracts along entire range, at up to 1300m.

HILL PIGEON *Columba rupestris* 33cm

A bluish-grey bird with two dark grey wingbars, glossy greenish-purple patches on breast and neck, whitish underparts and red legs. A distinctive broad white bar across the dark tail. Call a high-pitched, fast *gut-gut-gut*. Uses its black, slightly decurved bill to pick shoots, grain and seeds. A high-altitude resident in the Himalayas, at 3000–5000m in the summer; appearing in open fields, around barren rocks, cliffs and gorges, within moderate distance of human settlements; can also be found to 6000m, feeding on grain dispersed by pilgrims and travellers crossing passes; during winter, flocks descend to lower valleys.

SNOW PIGEON *Columba leuconota* 34cm

A typical pigeon, its coloration camouflaging it even at close range on gravel, soil or melting patches of snow. White and soft grey, with contrasting dark grey head, and distinct broad white bar (triangular in flight) across dark tail. Call a repeated croak. Quite tame except in remote areas, where whole flock takes off when disturbed, circling in amazing aerobatics and twists, in perfect unison, before settling. Eats seeds, grain and green food. High-altitude resident, roosting on rocky cliffs or feeding on grassy slopes, arid plateaux or around snowfields, all over the Himalayas from 4000m up to snowline; descends in winter to valleys, down to 1500m, forming large, restless flocks.

ORIENTAL TURTLE-DOVE *Streptopelia orientalis* 33cm

A large dove, reddish-brown with scaly-patterned back, two black 'chessboard'-like patches on each side of hindneck, and white terminal border to rounded tail. Fore area mainly vinous-rufous, fading to whitish chin, central throat and abdomen. Legs and bill red-brown. Call a typical *gorr gur-grugroo*. In display, rises to 30m or so with clapping wingbeats, then glides down in wide spiralling circles with fanned tail, landing on same branch. A ground-feeding, granivorous bird, preferring seeds and green shoots; also frequents camp sites to collect scattered grain. Fairly common resident of the lower wooded valleys of the Himalayas, to 4000m.

EMERALD DOVE *Chalcophaps indica* 27cm

Thakur Dalip Singh

A small forest-dwelling ground dove, larger than a myna. Wings and back dark shining green, and tail black with white outer feathers. Head, neck and underparts varying shades of grey, with white forehead and supercilium and a white bar on shoulders. Bill coral-red; legs pink. Flight very swift and silent. Active in morning and evening. Can be rather difficult to approach, but may be seen running about gleaning food on forest trails and paths, or near cultivated clearings in forest. Feeds on seeds, grain, fallen berries and fruits in forests and plantations, also in secondary growth. Resident at up to 1800m in the lower Himalayas, from Kashmir (Jammu) eastwards.

ASHY-HEADED GREEN-PIGEON *Treron phayrei* 28cm

Rupin Dang

A small, green, red-legged pigeon. Male has a dark grey crown and nape, chestnut-maroon back and scapulars, yellow in wings, black shoulders, and broad grey terminal tail band; bright yellowish-green throat and orange breast. Female lacks chestnut-maroon back and orange breast; has undertail-coverts buff, broadly streaked with green. Utters rich, whistling notes. Arboreal, in small flocks, often with other birds on fruiting trees. Feeds on fruits and berries. Inhabits forests, groves and orchards in the west-central Himalayas.

WEDGE-TAILED GREEN-PIGEON *Treron sphenura* 33cm

Rupin Dang

Has yellowish-green plumage and a wedge-shaped tail. Male has a rufous-orange crown, deep maroon on back and scapulars, with pale orangey breast. Female lacks rufous-orange on crown and maroon on upperbody. Gives rich, mellow, whistling calls; also *coo-coo* notes in summer. Mostly arboreal. Small flocks feed on foliage of fruiting trees, acrobatically reaching out to obtain fruit; occasionally feeds at salt-licks on ground. Found in mostly broadleaf hill-forests of the Himalayas, from Kashmir to extreme north-east India, at up to 2500m.

BLACK-NECKED CRANE *Grus nigricollis* 145cm

A rare crane with a very restricted distribution within our region. Tall and white, with head, neck and all flight feathers black, the last especially noticeable in flight; lores and crown naked and dull red; small white patch behind eye. Immature slightly smaller, with black areas duller, and head and neck brownish. Feeds on fallen grain, shoots, tubers, insects and possibly molluscs. Endemic to the Tibetan Plateau, at 4300–4600m, frequenting high-altitude marshes, lakesides, open cultivation. Can be heard trumpeting before migration in February– March, or later on the breeding grounds in Ladakh.

WATER RAIL *Rallus aquaticus* 28cm

Kamal Sahai

An unobtrusive bird, but commoner than it appears. Brown above, streaked black; white throat and grey breast, with flanks barred black and white. Long red bill diagnostic. Young have white barring on wings. Has a shrill squeal, sometimes heard early in morning. Mostly solitary, though mate is usually nearby. Secretive and cautious, moving with head held high, vanishing on slightest suspicion; legs dangle in flight. Feeds on molluscs, insects and shoots of marsh plants. Inhabits waterbodies. Breeds in Kashmir, spreading to the Indo-Gangetic plains in winter.

BAILLON'S CRAKE *Porzana pusilla* 18cm

Tim Loseby

A quail-like swamp bird, the smallest rail. Upperparts rufescent olive-brown, richly white-spotted; ash-grey cheeks and supercilium, with brownish streak through eye over eye-coverts to neck side; underparts brownish-grey to lower breast, then barred black and white to tip of undertail-coverts. Call a loud, high-pitched *crek*, followed by a pause and then another *crek*. Usually singly or in pairs, but cautious, heard more often than seen. Feeds on seeds of aquatic plants, also insects, worms, molluscs. Lives in reedy marshes, jheels and irrigated fields. Abundant breeder in Kashmir (1800m), and winters throughout the subcontinent; recorded as high as 4500m during migration.

TIBETAN SANDGROUSE *Syrrhaptes tibetanus* 48cm

A sandy-coloured, squat ground bird with long, pointed tail. Male has top of head white, finely barred black, mantle reddish sandy-brown, lower back and rump greyish-white with black vermiculations, and tail chestnut with white tips (elongated central feathers sandy-grey); face orangey, and underparts white, with breast finely barred black. Female has coarsely splotched mantle barred with black. Typical sandgrouse call, *cag-cag*; noisy in flight. Terrestrial, in flocks of 10–30, feeding on seeds and shoots; fairly tame. Comes to water early in morning and at dusk. On barren stony, semi-desert steppes on the Tibetan Plateau (eastern Ladakh, north Himachal and north Sikkim) at 4200–5400m; descends to lower elevations (locally) in winter.

SOLITARY SNIPE *Gallinago solitaria* 30cm

Joanna Van Gruisen

A cryptically coloured, plump-bodied mountain bird with a very long bill. Dense plumage brown, buff, black and fulvous with some chestnut markings above, but the various species of snipe are difficult to distinguish in field. Usually solitary, secretive and silent; mostly seen only when flushed, when it utters a fairly loud *pench* call in short, erratic flight. Feeds on small snails, worms and aquatic insects. Found in dense undergrowth along marshy mountain streams. Breeds at about 2800–4500m along the Himalayas, from east to west.

COMMON SNIPE *Gallinago gallinago* 27cm

Thakur Dalip Singh; Fotomedia

A marsh bird with cryptic plumage coloration. Brownish-buff above, heavily streaked and marked buff, rufous and black; dull white below. Has fast, zigzag flight; whitish wing-linings distinctive, but not easily seen. Several birds in dense marsh growth, very difficult to see unless flushed. Feeds mostly during morning and evening, often continuing through night; probes with long beak in soft soil, often in shallow water, for small molluscs, worms and insects. Occurs in wetlands, paddies and jheel edges. Breeds in parts of the west Himalayas at up to 2800m, and found higher up during migration.

IBISBILL *Ibidorhyncha struthersii* 40cm

Highly distinctive. Mainly ashy-brown above, with black crown, face, throat and breast band separated from grey upper breast by a narrow white bar; rest of underparts whitish. Most striking feature is the long (7–8cm), strongly decurved, dark red bill. Long red legs allow it to wade breast-deep in fast-flowing, icy streams, thrusting its bill under pebbles for water insects, crustaceans and molluscs. Flies between islets, often uttering loud *klew-ti-ti-ti-ti-klew*; may swim. Frequents shingle banks and islands in clear mountain rivers at up to 4000m; resident in extreme north-west India so long as rivers are ice-free, otherwise migrates in winter to Kashmir or towards the Himalayan foothills.

LESSER SAND PLOVER *Charadrius mongolus* 19cm

An attractive, gregarious wader. In summer plumage has rufous breast band, neck sides and crown, and prominent black eye-stripe and band over forehead (female paler, with black bands faint); otherwise sandy-brown above, white below. In non-breeding season becomes grey-brown and white. Whitish wingbar visible in straight, fast flight. Rather silent unless on territory, when utters a short repeated *pip-ip*; shrill *co-rup co-rup* alarm before taking wing. Feeds on small crabs and aquatic invertebrates. Summer visitor to high-altitude lakes on arid plains of the Himalayas, in Ladakh, Lahaul and Sikkim, between 3900m and 5500m; migrates in September.

BROWN-HEADED GULL
Chroicocephalus brunnicephalus 46cm

A medium-sized gull, the size of a Large-billed Crow. Mantle and back pale grey, with white rump, and underparts all white; wingtips black, with a white spot or 'mirror' on the outer primaries. In breeding plumage has a brown hood; at other times, head white with dark spot on ear-coverts. Red bill and legs. Floats buoyantly and high in water. A gregarious scavenger, associating with kites and other gulls; feeds on fish, insects, grubs and earthworms. Frequents large waterbodies. Breeds in Ladakh; migratory.

COMMON TERN *Sterna hirundo* 35cm

An elegant pale brownish-grey bird with white neck and underparts. In summer, has a black cap from forehead over crown to lower nape, extending down to eyes, a black-tipped red bill and red legs. Outer primaries black, best visible in flight, as is the deeply forked white tail, elongated outer feathers of which do not project beyond wingtips when at rest. In winter, cap is reduced to a mottled blackish patch on crown. Call a short *kik-kik*. Mainly hunts for fish, hovering over water and plunging headlong; also takes insects, molluscs, crustaceans. Fairly common visitor, May–September, at rivers and lakes at altitudes of up to 4500m.

OSPREY *Pandion haliaetus* 56cm

Gertrud and Helmut Denzau

A kite-sized raptor found near large waterbodies. Contrasting dark brown upperparts and pure white underparts and head, with dark brown breast band and dark band through eye. In overhead flight, wingtips and carpal patches darker, white underparts and breast band striking; tail squarish. Seen perched on a tree or tall post, or flying about, at times hovering to take a better look. Catches fish by plunging into water, the prey then taken to a perch to be devoured. A migratory raptor, with winter quarters throughout the subcontinent; small numbers breed in the Himalayas, between 2000m and 3300m.

ORIENTAL HONEY-BUZZARD *Pernis ptilorhyncus* 68cm

One of the few birds specialized in preying on bee colonies, especially on the large Rock Bee, this raptor has the head completely feathered, including the lores. It shows a short nuchal crest when seen in profile, and a smallish pigeon-like head. Plumage colour varies considerably: upperparts from buffish to almost black; underparts usually pale brown, barred with white; the tail is banded. Usually seen gliding in pairs or individually, or perched on a treetop. Found in areas with trees, from forest to cultivated land, from about 1800m in the Himalayas east to Assam and Bangladesh.

GREY-HEADED FISH-EAGLE *Ichthyophaga ichthyaetus* 74cm

Dark brown, with pure grey head, neck and throat, pale brown crown and nape, and white flanks, abdomen and tail, tail with broad black terminal band distinctive in overhead flight. Has a loud, ringing call; particularly noisy when breeding. Mostly solitary. Can be seen sitting upright on lookout perches, usually on trees over and around forest streams and pools. Feeds predominantly on fish, which it captures at the surface; does not plunge; sometimes also eats birds and squirrels. Resident, with considerable local movement depending on food supply, occurring around lakes and rivers in forested country, at up to about 1800m along the entire Himalayan range southwards.

LAMMERGEIER *Gypaetus barbatus* 125cm

Huge eagle-like vulture which, unlike others, has feathered head and neck. Dark brown and silver-grey above, streaked with white; pale rufous below; black 'beard' on whitish face gives it its alternative name of Bearded Vulture. Juvenile all dark brown with brown-black head, and whitish streaks on back. Legs fully feathered. Long, pointed wings (reaching 3m in span) and long, wedge-shaped tail allow it to glide effortlessly along cliffs and valleys, or soar high over mountains. Silent; occasional guttural hiss, especially in breeding season. Feeds entirely on bones; those too big to swallow are carried high up and dropped on rocks to shatter them, and marrow then eaten. Resident around cliffs and rocky areas at 5500m and above; descends to lower foothills in winter.

HIMALAYAN GRIFFON *Gyps himalayensis* 120cm

Huge, bulky, khaki-coloured vulture with sandy-white back, tawny below with extensive whitish streaks. Whitish-brown ruff around lower neck; upper neck almost naked, head sparsely covered with whitish down. Impressive in flight, circling high up or gliding along barren mountain slopes and cliffs, with wingspan almost 3m; underwings whitish, with almost black flight feathers with distinctly spread 'fingers', and short square tail dark brown. Immature dark brown with whitish shaft stripes. Large, heavy bill. Utters hoarse kakakaka, especially when fighting over carcass or at roost. After a heavy meal requires long run, beating its wings, until it takes off. A true mountain bird, resident in the Himalayas sometimes to above 5000m; descends to lower valleys in winter.

MONTAGU'S HARRIER *Circus pygargus* 46–48cm

An elegant raptor, slightly smaller than a kite. Male's upperparts ashy-grey, tinged brown, with grey throat and breast; rest of underparts white with chestnut streaks; black bar across secondaries diagnostic at rest and in flight. Female and immature variegated brown, with white rump and barred tail. Roosts at night in open, often in sizeable groups and with other harrier species. Feeds on frogs, lizards, mice and young ground-nesting birds. Can be seen at up to 2000m, gliding gracefully on spread, motionless wings over open country and meadows, and as high as 4500m during migration in spring and autumn.

CRESTED GOSHAWK *Accipiter trivirgatus* 40–46cm

Manjunath Hegade

Dark brown above, with slaty-grey crown and crest, and white below, streaked on breast and barred with rufous below breast. White throat and undertail-coverts; black throat stripe. Like Shikra (*Accipiter badius*), remains hidden in leafy branches, preferably around a forest clearing; has favoured hunting grounds; often soars over forest. Pounces on prey, especially medium-sized birds such as pigeons and partridges; also eats rodents and squirrels. Resident species of deciduous and semi-evergreen forests, in the Himalayas from Garhwal through Nepal, Sikkim and Bhutan, at up to 2000m.

BESRA *Accipiter virgatus* 29–34cm

A medium-sized, short-winged hawk. Male blackish-slaty above, with white nape-feather bases visible. Chin and throat white, with broad blackish throat stripe and two faint moustachial streaks; upper breast and flanks rufous, lower breast and abdomen barred. Square-ended tail is grey, with three or four blackish bands visible below. Female dark chocolate-brown above, becoming slaty-black on crown and nape. Very noisy while nesting. Perches on tall dead trees on edge of evergreen forests. Very quick on the wing in pursuit of prey, chiefly small birds, mice and bats. Resident in broken forested country along the Himalayas, at up to 3000m; descends to foothills in winter.

NORTHERN GOSHAWK *Accipiter gentilis* 49–60cm

Mohit Agarwal

A large hawk with a broad, distinctive eyebrow, and lacking crest and throat stripe. Strongly resembles the female Eurasian Sparrowhawk (*Accipiter nisus*) in coloration and physical appearance. Underparts of adult are white, finely barred pinkish-brown; upperparts are a uniform grey. Tail prominently and broadly barred black and white. Eyes deep red. Juvenile browner, with pale feather edges and bright yellow eyes. Has a powerful flight. Feeds on pigeons and other birds, as well as small mammals. Found in temperate and subalpine forests along the Himalayas, at up to 4800m.

COMMON BUZZARD *Buteo buteo* 50–55cm

Like other buzzards, occurs in light and dark plumage phases. Often creamy-brown head and patchy underwing pattern; identification confusing, however, and reliable separation from other buzzard species difficult. All have a mewing, whistle-like call. Seen singly, soaring over treetops and open countryside; rarely hovers, but hangs motionless in wind. Perches for long periods in leafy trees. Feeds on rodents, young birds, snakes and worms. Fairly common in cultivated fields, forests and nearby rivers in Kashmir, Garhwal, Nepal and Sikkim, at up to 2500m; higher during spring and autumn migration.

LONG-LEGGED BUZZARD *Buteo rufinus* 55–60cm

Extremely variable in coloration, ranging from blackish-brown to red-brown and fulvous to pale sandy. Head, neck and breast vary from brown to almost white. Very vocal during the breeding season. Usually seen in pairs, perched on a treetop, rock or mound. Groups sometimes gather at forest fires, feasting on lizards, rats and insects in company with other raptors and drongos; also eats small mammals, reptiles and frogs. Hunts live prey by pouncing from a perch; also from about 30m up in the air, where it hovers to scan the ground. Breeds in the north-western Himalayas, at up to 4300m.

GREATER SPOTTED EAGLE *Aquila clanga* 64–70cm

Kamal Sahai

A larger eagle, deep blackish-brown above with purplish wash on back, somewhat paler below, and often with whitish rump. Rounded nostrils. Immatures may have white markings above. Utters a loud, shrill *kaek kaek*, often from perch. Soars on straight wings with drooping tips. Mostly solitary, and sluggish, perching for long spells on bare trees or on ground. Feeds on small animals, waterfowl and small birds. Frequents tree-covered areas, preferably in vicinity of water. Breeds sporadically in parts of north India at up to 1500m, but migrating birds can be observed in spring and autumn at up to 4500m.

STEPPE EAGLE *Aquila nipalensis* 75–80cm

Plumage coloration variable, from dark blackish-brown to pale, almost buffish-brown, and often with a rufous nape patch. Has oval, elongated nostrils. Immatures usually show two diagnostic pale bars on the upperside and underside of wings. Has a high-pitched call. Lazy, often low flight. Often seen on ground, solitarily or several scattered birds, eating carrion or offal; also feeds on small animals and birds. Winter visitor to Nepal and north India, frequenting open treeless country, in vicinity of habitation and cultivation.

GOLDEN EAGLE *Aquila chrysaetos* 90–100cm

A large eagle, dark brown with tawny neck, powerful bill, and yellow cere and feet. Younger immatures have white tail with broad black terminal band and conspicuous white patches at centre of broad wings, visible when flying overhead; immatures seen more often than adults. Typical flight, with wings held in shallow V-shape while soaring. Utters thin, piercing yelp. Hunts in pairs, stooping at high speed on prey; feeds mainly on mammals and birds (even Demoiselle Cranes *Anthropoides virgo*). Resident of high mountains with crags, precipices and sparse vegetation all over the Himalayas, to 5000m and above in summer, moving to lower valleys in winter.

BOOTED EAGLE *Aquila pennata* 50–54cm

Our smallest eagle, with two distinct colour phases. Light phase has paler head, uppertail and upperwing-coverts; buffy-white wing-linings, underbody and tail with blackish flight feathers distinctive in flight. Dark phase chocolate-brown below, with paler, banded tail. Has loud scream of several notes. Single birds or pairs can be seen hunting in concert. Picks prey off ground or chases it in flight; feeds on birds, rodents, lizards; steals poultry. Found in open forests, scrub orchards and by water, also around human habitation. Breeds in the Himalayas at 1800–3000m, sporadically in the peninsula; commoner in winter throughout region.

CHANGEABLE HAWK-EAGLE *Nisaetus limnaeetus* 72cm

A slim, powerful, kite-sized raptor. Brown plumage, streaked with white below, the streaks fine on throat and becoming thicker and bolder on breast. Black saggital crest and long feathered legs distinctive. Loud high-pitched call, ending in a scream; heard more often in breeding season. More of a large hawk than an eagle, sweeping down and ambushing prey from high perch on top of a well-foliaged tree, from where keeps keen lookout. Preys on hares, large birds, also squirrels, lizards and the like. Resident of forested habitats in the sub-Himalayan terai and duars, from Garhwal eastwards, in the foothills and up to 1900m.

COMMON KESTREL *Falco tinnunculus* 36cm

A dove-sized falcon of open country. Head grey, streaked darker, with characteristic dark moustache. Black-spotted rufous above, with grey rump, uppertail-coverts and tail, tail white-tipped with black subterminal band; pale rufous-buff below, breast and flanks streaked and spotted with brown. Female barred above, and with bolder markings then male below; tail barred, with white tip and dark subterminal band. Both sexes have wingtips darker. Well known for its habit of hovering; feeds on insects, lizards and small rodents, captured by parachuting down. Breeds in the west Himalayas, at 700–4300m; in non-breeding period, common above treeline to 5500m.

MERLIN *Falco columbarius* 31–36cm

A dove-sized raptor. Head greyish, thinly streaked dark, with whitish supercilium; pale blue-grey above, with broad rufous and black hindcollar, tail with white tip and broad subterminal black band; white below, tinged rufous, and boldly streaked blackish, with thighs and undertail-coverts darker rufous. Female larger and browner than male, tail barred throughout. Hunts singly in open areas, particularly among crops, perching on ground or mounds separating fields. Flies very fast and low, with short, clipped wingbeats and glides on half-closed wings. Captures small birds by swift, direct pursuit. Seen more often in Ladakh than in the more western Himalayas.

EURASIAN HOBBY *Falco subbuteo* 31–36cm

A dove-sized falcon. Sexes alike, slaty-grey above with darker head and moustache, and streaked below with a red undertail; immatures have underparts heavily dark-streaked. Calls and general habits are similar to those of other falcons, but both diurnal and crepuscular. Flies in wide arcs, rising and falling in height, attacking aerial prey in a swift downward swoop; feeds on small birds, small bats and large flying insects, pursued, captured and devoured on the wing. Resident of open-wooded country and tree-savanna type of habitat in the lower Himalayas, between 1800m and 2400m.

PEREGRINE FALCON *Falco peregrinus* 40–48cm

Size of the Large-billed Crow, and a typical streamlined falcon. Dark ashy-grey above and fulvous-white below; conspicuous black moustache and finely dark-barred underparts. Has a fairly loud, ringing scream. Usually found individually. Flight direct and powerful, a few fast wingbeats followed by a glide. Like other falcons, spends a lot of time waiting for prey from a vantage perch. Captures prey in a fast stoop, raking with its hindclaw; eats a wide variety of bird species, but especially waterfowl. Found near wetlands along the Himalayas, at up to 2200m, and higher during its migration in spring and autumn.

Kamal Sahai

GREAT CRESTED GREBE *Podiceps cristatus* 50cm

The largest grebe in the Indian region, much larger than the Little Grebe (*Tachybaptus ruficollis*), and like them almost tailless, with legs placed far back. Dark greyish-brown above and whitish below, with a long slender white neck; two backward-directed ear tufts, smaller on the female, and largely absent in winter. White on wing prominent in flight. Dives without a splash; when alarmed, has a habit of swimming underwater, surfacing, taking a look, and repeating the process until sufficient distance gained. Winter visitor in small numbers in northern India, scattered on reed-fringed or vegetation-covered fresh and littoral waters.

LITTLE BITTERN *Ixobrychus minutus* 35–36cm

Gen. R.K. Gaur

A small bittern, black above, ochre and maroon on nape and below. Large whitish patch on dark wings, particularly striking in flight. Juveniles heavily streaked brown above and below. Usually solitary, and may be seen among aquatic vegetation at dawn and dusk. Very active during the rains, when it also breeds; courting male gives *kok* calls. Feeds on insects and frogs, also fish. Commonly seen in the Kashmir Valley, at up to 1800m, in marshy areas, dense reed growth and paddy cultivation.

CINNAMON BITTERN *Ixobrychus cinnamomeus* 38cm

Kamal Sahai

A nocturnal bird, reddish-brown above and much paler below. Female differs from the male in having the upperparts mottled with dark and light spots, and underparts which are streaked darker. Spends the day quietly concealed in a reedbed, only flying out at dusk to feed; usually seen during day only when flushed from roosting sites. Not at all gregarious. General behaviour very much like that of the common Indian Pond-heron (*Ardeola grayii*). Found all over India, especially in the south of the Himalayan terai, inhabiting reedbeds and well-vegetated inundated paddy cultivation.

ASIAN FAIRY-BLUEBIRD *Irena puella* 27cm

Named after the Greek goddess of peace, Eirene, and *puella*, meaning a maiden, thus suggesting something pretty, the Asian Fairy-bluebird is indeed a very beautiful bird. A brilliant, shimmering deep blue above and velvety-black below, the male is one of the most striking birds of our forests. The female is more plain, being dull greenish-blue with dark lores. Has a rich and mellow *peepyt* call, uttered frequently. Pairs or small groups found in canopy trees, feeding on fruits and nectar, sometimes coming lower down into the understory. Inhabits wet forests of the eastern Himalayas east of extreme southeast Nepal.

GOLDEN-FRONTED LEAFBIRD *Chloropsis aurifrons* 19cm

A.V. Manoj

Well camouflaged in its leaf-green plumage, this species is more easily heard than seen, but often glimpsed when flying from canopy of one tree to another. Golden-orange forehead, blue shoulder patches, dark blue chin and throat (tending to be darker in the south) and a black border to the throat are characteristic. Female duller, with smaller orange patch on forehead. An excellent mimic of other birds, often putting to serious test one's capabilities for recognizing calls. A canopy species feeding on fruit, nectar and insects, and a frequent member of the mixed foraging flocks of the wet forests. Resident of well-wooded areas in the Himalayas, at up to 1800m.

ORANGE-BELLIED LEAFBIRD *Chloropsis hardwickii* 19cm

An elegant bird, grass-green above, with chin, throat and upper breast velvety-black with a purplish sheen, and underparts deep orange. On the female, the blue moustachial streak is less bright and the underparts a paler orange. Slender bill is black and slightly curved. Noisy, and an excellent mimic; a mix of its own notes and imitations of other birds' calls, notably of drongos, bulbuls, shrikes and cuckooshrikes. Restless, and purely arboreal, pairs or parties seen in leafy trees; feeds on insects, flower nectar and fruits. An important agent of pollination. Found in light forests and gardens at 600–2500m along the Himalayas.

Rupin Dang

GREY-BACKED SHRIKE *Lanius tephronotus* 25cm

A medium-sized shrike, dark grey above, with a broad black band from lores through eyes towards nape. Cheeks, throat and upper breast dirty white with rufous tinge, becoming fulvous towards vent, rump and uppertail-coverts; wings and longish tail dark brown. Strong blackish bill, with upper mandible clearly hook-tipped, overlapping the lower mandible. Harsh *ktacht-ktacht-ktacht* call is given mainly after sunset or when alarmed. Feeds on larger insects (grasshoppers, crickets), also lizards and small rodents. Resident in dry, open scrub and thorny bushes from the western Himalayas eastwards to Burma, at up to 4500m; winters in the foothills.

EURASIAN JAY *Garrulus glandarius* 31cm

A noisy, aggressive bird. Has a pinkish-brown plumage, with velvet-black malar stripe, closely black-barred blue wings, and white rump contrasting with a jet-black tail. It utters screeching notes and whistles, also a series of guttural chuckles; a good mimic. Small parties often accompany other Himalayan birds. Mostly keeps to trees, but also drops down into lower bushes and to the ground. Feeds mainly on insects, fruits and nuts. Found in mixed temperate forests along the Himalayan range from west to east, between about 1500m and 2800m; descends in the winter.

BLACK-HEADED JAY *Garrulus lanceolatus* 33cm

A typical jay, inquisitive and aggressive, and with a laboured flight. Has a black cap, black and white face and white in wings. Gives guttural chuckles, screeching notes and whistles; a good mimic. Small noisy bands often accompany other Himalayan birds. Prefers cover of trees, but also descends into bushes and to ground. Feeds on insects, fruits and nuts. Found in mixed temperate forests of the west Himalayas, east to about central Nepal; common and familiar about Himalayan hill stations, at up to 3000m.

59

YELLOW-BILLED BLUE MAGPIE *Urocissa flavirostris*
66cm including tail

A pigeon-sized, beautiful bird with long graduated tail and streamers. Head, neck and breast black; nape patch and lower underparts white; rest of the body an eyecatching purplish-blue. Bill, legs and feet orange-yellow. A bird with a very wide vocabulary, frequently mimicking other species. Gregarious in non-breeding season, keeping together in small flocks. Mainly tree-dwelling, but does come down to shrubs and ground. Feeds on a variety of foods, e.g. fruits, insects and other animal matter, also grain. Resident of the Himalayan wet temperate mixed forests.

RED-BILLED BLUE MAGPIE *Urocissa erythrorhyncha*
70cm including tail

Rather similar to the Yellow-billed Blue Magpie in plumage and habits. Pigeon-sized, with the head, neck and breast black, a long graduated tail with extended streamers, and off-white underparts; white patch on nape and hindneck larger than on Yellow-billed. Bill, legs and feet red. Noisy; calls a mix of metallic screams, loud whistles and raucous notes. Shares habits and habitats with the Yellow-billed species, but tends to prefer lower elevations. Also tends to be more confiding, therefore appearing more frequently at the Himalayan hill stations.

GREY TREEPIE *Dendrocitta formosae* 40cm

A typical treepie, long-tailed and slender. Plumage grey and sooty-brown, with the elongated central rectrices (blunt and rather spatulate) ashy-grey and broadly black-tipped. Has varied, harsh and raucous calls. Keeps in small parties of four or five birds or in loose flocks in tall forests and tea gardens. Descends to lower branches to feed on fruit, seeds, flower nectar, insects, lizards, eggs and young of small birds. Resident of broadleaf forests in the Himalayas, between 600m and 2600m; subject to altitudinal movements, coming lower in winter.

BLACK-BILLED MAGPIE *Pica pica* 52cm including tail

A handsome black and white bird with long tail. Wing-coverts and tail glossy dark blue-green in direct sunlight; scapulars and most of underparts pure white. Strong, slightly decurved, longish black bill. Call a hoarse *querk*. Prefers open habitats around villages, foraging in fields or sitting on rocks, trees or rooftops; hops or stalks on ground in crow-like manner. Takes anything edible: insects, small mammals, young birds, eggs, grain, fruit, kitchen scraps. When disturbed, takes off in fast, straight, flapping flight, long tail trailing, uttering harsh *kekk-kekk-kekk* alarm. Resident in valleys and around settlements mainly in the trans-Himalayan areas and Himachal, at up to 4500m.

HUME'S GROUNDPECKER *Pseudopodoces humilis* 20cm

A small, active, grey-sandy bird with slightly darker wings. Whitish outer tail contrasts with dark brownish centre, visible in fluttering flight; broad neck collar, cheeks and entire underparts cream, while darker stripe from upper nape through eye to slender, decurved blackish bill is conspicuous feature. Call a high-pitched *psiii* or a fast *cheep-cheep-cheep*. Mainly terrestrial, moving in long hops. Pecks hoopoe-like in the soil for insects. Resident of high-altitude arid plains and open dry gullies in eastern Ladakh and northern Sikkim, to 5500m, descending lower in winter (seen more often in neighbouring Tibet, even in villages).

RED-BILLED CHOUGH *Pyrrhocorax pyrrhocorax* 45cm

An all-black bird, the size of a House Crow (*Corvus splendens*), with bright red legs and slender, slightly curved red beak. Has a musical plaintive call, as well as other calls. Gregarious, forming large flocks, sometimes of over a hundred birds. Confiding; roosts in man-made structures, rock fissures and caves. Quite tame near villages, grubbing in fields. Feeds on grubs, insects, barley; not given to scavenging so much as the other members of the crow family. Frequents high mountainous areas with cliffs, alpine meadows or cultivation, in the Himalayas between 2200m and 5400m; may descend to about 1500m in severe winters.

YELLOW-BILLED CHOUGH *Pyrrhocorax graculus* 39cm

S.G. Neginhal: Sanctuary Photo Library

A glossy jet-black crow-like bird with yellow bill and bright red legs. Has a high-pitched, musical *guee-ah* or *cree-ah*. Gregarious and sociable, keeping in family parties, occasionally fairly large flocks, throughout year. Seen in alpine meadows and pastures, digging for grubs and insects; takes food scraps around human camps and settlements. High-altitude resident in the western Himalayas, found in moist and dry temperate mountains, alpine meadows, pastures and cultivation, at up to 5500m and above; tolerant of cold, remaining at high elevations even in winter.

EURASIAN JACKDAW *Corvus monedula* 33cm

A small, slaty-black crow, distinguished from House Crow (*Corvus splendens*) by a broad silvery-grey hindcollar, rather short thick neck and greyish-white eyes. Call *chack* or *jack* or *kwai*, more musical than other crows. Gregarious, sociable and inquisitive, usually in huge flocks in association with other crows, feeding in damp pastures and flooded meadows. May be seen on backs of grazing sheep and cows, searching for insects and larvae; also eats slugs, fruit and grain. Can be tame or shy, depending on conditions. Breeds abundantly all over the north-western Himalayas; after breeding, disperses to higher pastures above treeline.

LARGE-BILLED CROW *Corvus japonensis* 50cm

An all-black crow, larger than the common House Crow (*Corvus splendens*). Plumage shiny and iridescent, the shades varying between regions and races. Produces a variety of gurgling croaks. A territorial, aggressive and daring species, not hesitating to attack larger animals and predators either to steal food or in defence of its young. Diet variable; will eat almost anything man would, and more! Roosts communally with other crows and mynas at sites that tend to become traditional, the same tree being used for decades. Groups collect to 'mourn' others of their species which have died. Appears at up to 3500m.

COMMON RAVEN *Corvus corax* 70cm

The largest crow species, all black, with a massive, heavy bill. In overhead flight, the conspicuous large, pointed, wedge-shaped tail is distinctive. Skilled flier, gliding on outstretched wings over long distances or indulging in spectacular aerobatics. Has an unmistakable *corax-corax* call. May be seen with vultures at carcases; omnivorous, it eats anything, dead or alive, and regurgitates pellets of undigested hair and bones, as raptors and owls do. Resident, preferring arid plains and rocky deserts high above treeline at 4000–5500m, and has been recorded at 6400m by Everest expeditions; descends to lower altitudes during the harsh winter.

BLACK-NAPED ORIOLE *Oriolus chinensis* 26cm

Rupin Dang; Fotomedia

A bright yellow bird with a prominent black mask meeting around the nape and a pink bill. Black wings and tail. The female is a dull yellow-greenish and slightly streaked, mainly on the underparts and especially the belly. Has a clear musical call. Occurs either singly or in pairs, often in association with other forest birds, but is difficult to see because it remains in the tree canopy. Feeds on insects and fruits. Found in forests, orchards and gardens in the central and eastern Himalayas, at up to 2300m.

BLACK-HOODED ORIOLE *Oriolus xanthornus* 25cm

Krupakar-Senani; Sanctuary Photo Library

Overall golden-yellow, with black head, black flight feathers and black central tail. Deep reddish-pink beak, and darkish legs. Sexes largely alike. Young have much less black on head and body, but breast is streaked with black. Call a nasal *kwaak*, a harsh *cheeah*, or melodious fluty whistles rendered as *tu-yow-yow or tu-yow*. Habits similar to those of the more familiar Golden Oriole (*Oriolus oriolus*). Usually in pairs or small groups, keeping to trees; rarely descends into lower branches or ground; active and lively; associates with other birds in mixed parties. Resident oriole of the moister areas, found at up to 1200m.

MAROON ORIOLE *Oriolus traillii* 27cm

Rupin Dang

A myna-sized oriole, glossy crimson-maroon, with head, neck
and wings black and tail chestnut-maroon. Female like male, but
underparts greyish-white, streaked blackish. Has a harsh *kee-ah*
followed by rich, melodious whistles, similar to Golden Oriole's
(*Oriolus oriolus*). Found either singly or in pairs in treetops in
dense forest, commonly in association with drongos, minivets and
nuthatches. Feeds on wild figs, berries, insects and nectar. Resident
of moist deciduous and evergreen forests from Himachal east
along the entire Himalayan range, at up to 2400m.

LARGE CUCKOOSHRIKE *Coracina macei* 28cm

Male grey above, with broad
dark stripe through eyes to ear-
coverts, black wings and tail,
and whitish below with greyish
breast. Female is barred grey
and white below, with paler eye-
stripe. Gives ringing whistle,
ti-treeeeee, the second note
long-drawn and higher, not unlike
Plum-headed Parakeet's call; very
active and noisy when breeding
(March–June). In pairs or small
bands in upper branches, but
may descend into bushes. Flight
over forest characteristic, with
few wingbeats and a glide, often
calling; flicks wings on landing.
Feeds on insects, larvae, nectar
and fruit. Forests, gardens and
tree-dotted cultivation through
most of India, to about 2200m in
the Himalayas; absent in much of
north-western India.

LONG-TAILED MINIVET *Pericrocotus ethologus* 18cm

A conspicuous, arboreal, slim, bulbul-sized bird, glossy black and deep scarlet in colour, with a large scarlet patch on black wing (a distinct identification pattern), a black throat, and black and scarlet graduated tail. Female has yellow instead of scarlet in plumage. Very gregarious except when breeding. Parties of 20–40 can be seen as they fly between treetops or hover in front of flowers. Occasionally catches insects in the air. Feeds on spiders, beetles, various larvae, fruit, as well as acacia buds. Found in open forest along the entire Himalayan range, at up to 3900m.

SCARLET MINIVET *Pericrocotus speciosus* 20cm

Male   Female

A flock of Scarlet Minivets illuminated by the rich red of the setting sun over the forest canopy is a spectacular sight. Male bright orange to scarlet (southern race orange) and black; female and immatures grey, bright yellow and black. On the male, the head and shoulders, wings and central tail feathers are black, while the female has only wings and central tail feathers black, the head and back being grey. Two bold bars running along, not across, the closed wing, orange-scarlet or yellow depending on sex. Keeps to trees in forested tracts. Resident throughout the lower Himalayas, at up to 2700m.

BAR-WINGED FLYCATCHER-SHRIKE *Hemipus picatus* 14cm

A small, black, flycatcher-like bird. Male has glossy black head and back, underparts pure white, a white collar around hindneck, white rump, and black and white wings and tail. Black replaced by sooty-brown in female. It utters a continuous squeaky call. Has a hunchbacked posture while perched on a branch. Moves from perch to perch, and makes short flights into the air to catch insects. Usually part of mixed foraging parties of small insectivorous birds. Found in scrub and dry and moist deciduous and evergreen forest in the Himalayas from Himachal Pradesh eastwards, at up to 1800m.

A.V. Manoj

YELLOW-BELLIED FANTAIL *Chelidorhynx hypoxantha* 9cm

A restless, diminutive flycatcher. Male dark, greyish-olive, above, with forehead and supercilium yellow, a broad black band from lores through eye and ear-coverts, and brown tail with conspicuous white shafts and tips; bright yellow below. Blackish olive-brown eye-band on female. Seen in lower canopy and taller shrubs, singly or in pairs, usually part of mixed parties of small insectivorous birds. Dances incessantly with fanned tail and partly drooping wings, constantly uttering distinctive *sip sip* call. Hunts for tiny winged insects from low bushes or canopy of tall trees. Found from the western Himalayas eastwards, at up to 4000m.

WHITE-THROATED FANTAIL *Rhipidura albicollis* 18cm

An active, lively bird with a slaty-brown plumage, including underbody. Has a white supercilium and throat and white tips to all but the central tail feathers. Sexes alike. Seen either singly or in pairs, flitting about in the low and middle levels of the vegetation. Fans tail, flicks wings, or bursts into a whistling trill. Makes short dashes up into the air to hunt insects. Inhabits light forests, groves and gardens among habitation and scrub along the Himalayan range, at up to about 3000m.

WHITE-BROWED FANTAIL *Rhipidura aureola* 17cm

Belongs to a very distinctive-looking group of flycatchers. A black, grey-brown and white bird with a prominent fan-like tail. Sexes similar, but the female is browner. Head and throat black and upperparts ashy-brown, with broad white bands extending back on each side of the forehead and joining at the rear to give a white nape; underparts white. Tail brownish, all but the central feathers tipped and edged white, the white increasing progressively outwards. Resident flycatcher, found in groves of trees in open country and cultivation, in gardens and in lightly wooded areas.

Mohit Agarwal

ASHY DRONGO *Dicrurus leucophaeus* 30cm

A dark bird with deeply forked tail. Upperparts are slate-grey with a strong blue-grey gloss; uniform dark grey underparts, with blue-grey gloss on sides of neck and breast. Noisy, having varied repertoire of harsh calls and pleasant whistles; a good mimic. Mainly arboreal and insectivorous, catching larger winged insects in flycatcher fashion by aerial sallies from perch. Very agile, especially when in pursuit of insects. Behaviour similar to other drongos. Frequents forests and wooded habitats. Locally common, breeding in the Himalayas from north Pakistan east to the north-eastern hills, from the foothills up to 3000m; winters in the plains.

HAIR-CRESTED DRONGO *Dicrurus hottentottus* 31cm

Rupin Dang

A strictly arboreal bird with glistening blue-black plumage, with fine hair-like feathers on forehead. Tail diagnostic: square-cut, and inward-bent (curling) towards outer ends. Longish, decurved, pointed bill. Noisy, with whistling, metallic calls and harsh screams. Solitary or in loose pairs; small numbers may gather on favoured flowering trees such as *Erythrina*, *Salmalia*, *Bombax*. Rather aggressive, driving away other birds; often in mixed foraging parties. Feeds chiefly on flower nectar; also insects, more so when rearing nestlings. Found in forested regions of the lower Himalayan foothills, east of Kumaon to north-eastern India, at up to 2000m.

GREATER RACKET-TAILED DRONGO
Dicrurus paradiseus 35cm

A large drongo, all black with a unique tail and a crest of backward-bent feathers. Extremely long shafts of outer tail feathers have central portion bare, with an expanded racket-like vane at the tip. Excellent mimic: has a variety of screams, whistles and perfect imitations of over a dozen species. Bold and aggressive, seen mobbing bigger birds 100m over forest. Shares the habits of the group in making sallying flights to catch insects. Perhaps spends more time on the wing than its congeners. Also takes flower nectar, being easily attracted to such blooms as *Bombax* and *Erythrina*. Inhabits deciduous and evergreen forests at up to 1500m.

Krupakar-Senani

BLACK-NAPED MONARCH *Hypothymis azurea* 16cm

Male is lilac-blue, with black patch on nape and gorget on breast, slight black scaly markings on crown, sooty wings and tail, and white belly. Female ashy-blue, duller, lacks black on nape and breast. Calls often, indicating its presence: has sharp, grating, high-pitched *chwich-chweech* or *chwae-chweech...*, slightly interrogative and delivered rapidly, uttered when on the move; short rambling notes when breeding. Solitary or in pairs, often in mixed foraging parties. Extremely active, flits and flutters about, often fans tail slightly. Feeds on insects. Found in forests, bamboo and gardens along the Himalayan foothills, to about 1200m, east of Dehra Dun.

Parvesh Pandya; Sanctuary Photo Library

ASIAN PARADISE-FLYCATCHER *Terpsiphone paradisi* 50cm

Male (left); female (above)

One of the most distinctive flycatchers and cannot be confused with any other. Old males are pure white, with head, crest and throat black, and the central pair of tail feathers elongated into extremely long (30cm) ribbon-like streamers. Female has white of upperparts replaced by rufous, dirty white underparts, and no tail-streamers. Very young males (1 year) resemble female (though with head and throat entirely black), the tail lengthening with age, but the rufous remaining until around the third year. Has a harsh wich call, and a rarely heard soft warbling song. Found in shady tree-covered habitats along the Himalayas from west to east.

WHITE-THROATED DIPPER *Cinclus cinclus* 20cm

As birds which plunge into hill-stream torrents, dippers form a unique group. This species has head and mantle chocolate-brown, rest of upperparts dark brown, and lower underparts chocolate-brown with contrasting pure white throat and breast. Has a harsh *dzchit-dzchit* call. Dense thick plumage effectively keeps away water when the bird dives among rocks, moving underwater using its wings and even walking on the bottom; feeds on aquatic insects and their larvae. Tends to fly low over the water. Found in rocky, icy, fast-flowing torrents and glacial lakes in the Himalayas, at up to 5000m.

BROWN DIPPER *Cinclus pallasii* 20cm

Entirely chocolate-brown, with prominent white eye-ring. Noisy: shrill *dzchit dzcheet* call, given mostly in swift flight low over water, audible above roaring torrent and usually heard long before bird is sighted; has a lively piercing song. Solitary or in pairs along a stretch of gushing icy stream. Settles on slippery rocks amid water; very restless and energetic, bobs and moves body from side to side. Plunges and swims against current, also walking on bottom; feeds on aquatic insects. A bird of rocky streams and glacial lakes in the Himalayas, between 2500m and 4500m, sometimes down to 2000m in winter.

BLUE-CAPPED ROCK-THRUSH *Monticola cinclorhynchus* 18cm

An elusive forest bird. Male has a blue crown and nape, black back, broad stripe through eye to ear-coverts, blue throat and shoulder patch, and chestnut underparts; white wing patch and chestnut rump distinctive. Female unmarked olive-brown above, buffy-white below, thickly speckled with dark brown. Breeding male has rich song; mostly silent in winter. Either solitary or in pairs, moving through the foliage in mixed parties or rummaging on ground among leaf litter. Feeds on insects, flower nectar and berries. Best seen when it emerges in clearings. Breeds along the entire range of the Himalayas, between 1000m and 3000m, sometimes higher.

CHESTNUT-BELLIED ROCK-THRUSH *Monticola rufiventris*
23cm

Male (left); female (above)

Rupin Dang

A myna-sized bird. Male brilliant cobalt-blue with some blackish on mantle, and black lores, ear-coverts and neck sides; throat blackish-blue, rest of underparts chestnut. Female has back and rump olive-brown with dark crescent-shaped bars, eye-ring and patch on neck side buff, ear-coverts dark grey-brown, and underparts scaled dark brown and buff. Has harsh indrawn rattle. Single birds or pairs perch upright on bushes and sway their tail slowly up and down. Feeds on the ground on beetles and grasshoppers; also catches winged insects like a drongo. Inhabits pine, oak, fir and deodar forests along the Himalayan range, at up to 3500m.

BLUE WHISTLING-THRUSH *Myophonus caeruleus* 33cm

Though not so fine a songster as its southern counterpart, the Malabar Whistling-thrush (*Myophoneus horsfieldii*), this species, too, has a whistling song which follows a fixed pattern with some variation, very human in quality, but clearer and more resonant. Pigeon-sized, and dark purplish-blue, spotted with brighter blue; forehead, shoulders, wings and tail also have brighter blue in them, and wings show some white spots. Bill yellow. In common with the group, has a loud, penetrating kreee alarm call. Found near torrential streams and gorges in heavy forest along the Himalayas, at up to 4000m; at times even higher.

PLAIN-BACKED THRUSH *Zoothera mollissima* 27cm

Tim Loseby

A rather short-tailed thrush, lacking wingbars. Has a loud rattling alarm note. Occurs in pairs, or several together in winter. Usually difficult to spot until it takes off from somewhere close by; flies into branches if disturbed. Feeds on ground, on insects and snails. A Himalayan bird, breeding around and above the timberline, with a preference for dwarf forests, scrub-covered rocky slopes and grassy mountainsides; found in heavy forest during winter, when descends to about 1000m.

SCALY THRUSH *Zoothera dauma* 25cm

Göran Ekström

A well-patterned thrush, with distinctly spotted back and boldly marked underparts. Mostly silent. Seen in pairs, or several together in winter. As is typical for this family, it feeds on the ground on insects and snails, difficult to spot until it suddenly takes off to fly up into branches. A bird of the Himalayas, east of central Himachal, breeding in timberline forest and scrub, at 2000–3600m; descends to the foothills and eastern plains in winter, when often found in heavy forests.

TICKELL'S THRUSH *Turdus unicolor* 21cm

Male is light ashy-grey, duller on breast and whiter on belly, with rufous underwing-coverts in flight. Female olive-brown above; white throat, streaked on sides, tawny flanks and white belly. Has a rich song, two-note alarm, also some chattering calls. Small flocks visit orchards and meadows with sparse undergrowth, sometimes with other thrushes. Flies into trees when approached too close. Hops fast on ground, stopping abruptly as if to check underground activity. Digs for worms; also eats insects and small fruits. Breeds in open forests and groves, in the Himalayas east to central Nepal and perhaps Sikkim, at 1500–2200m; winters along foothills east of Kangra.

GREY-WINGED BLACKBIRD *Turdus boulboul* 28cm

Male has black plumage with diagnostic large grey wing patch. Yellow eye-ring and orange beak. Female olive-brown, with brown wing patch and a duller bill. Gives distinct chuckling notes; has a rich, loud and fluty song, and guttural *churrrr* note. Solitary or in small parties, sometimes with other thrushes in winter. Shy, taking to trees on slightest suspicion. Feeds on ground; picks insects, digs for worms, plucks small fruits. Breeds in Himalayan broadleaf forests, also scrub, secondary growth and near rural habitation, at 1800–2700m; descends to about 1200m in winter, sometimes into foothills and adjoining plains.

BLACK-THROATED THRUSH *Turdus atrogularis* 25cm

Toby Sinclair

Male grey-brown above and white below, with black lores, sides of neck, throat and breast. Black area scaled white in winter. Rufous replaces black in Red-throated Thrush (*Turdus ruficollis*). Female browner above; white below, streaked blackish on sides of throat, with dull ashy-brown breast and flanks, streaked brown. Harsh *schwee* call; chuckling *wheech-which*. Gregarious, often with other thrushes. Hops on ground, flying into branches when disturbed. Feeds on invertebrates and fruit. Common winter visitor to Himalayan foothills, to about 4500m on spring and autumn migration, occurring in forest edges, cultivation, scrub and fallow land.

MISTLE THRUSH *Turdus viscivorus* 29cm

Slightly larger than a myna, both sexes are grey-brown above, with buff underparts boldly marked with roundish dark brown spots. Flight feathers more or less edged with white; pale eye-ring. The white underwing-coverts are conspicuous in flight; outer rectrices tipped with white. Has a loud song, generally delivered from treetops, and a rattling alarm. Seen either singly or in pairs during breeding season, thereafter commonly in flocks of 20 or more. Moves on the ground in long hops, in search of insects, larvae and berries. Flies into trees when disturbed. An open-forest bird, found along the Himalayas east to west Nepal.

GOULD'S SHORTWING *Brachypteryx stellata* 13cm

Peter Morris

A small, dumpy bird with characteristic short tail. Chestnut above, with blackish lores; finely barred grey and black below, with diagnostic triangular spots on lower breast and belly, rufous wash on flanks and vent. Usually silent; occasional *tik-tik* call. Mostly solitary. Generally keeps to ground, among dense undergrowth in evergreen hill-forest, but occasionally ascends to bushtops and low branches; shy, but may allow close approach in some localities. Feeds on insects and grubs. Breeds in the Himalayas, east of Kumaon, in dense forest growth and also among boulders in alpine country, at 3000–4200m; descends in winter to 1500m.

DARK-SIDED FLYCATCHER *Muscicapa sibirica* 13cm

A dark, sooty-brown flycatcher with noticeably large eyes and a pale eye-ring. Grey-brown below, with a whitish throat patch and centre of belly. Generally silent, though has a high-pitched, thin and reedy song. Fond of forest glades and clearings littered with tangled brushwood and tree stumps, where perches upright. Catches insects by making aerial sallies, returning to same perch. Found in open forests of pine, deodar, fir, birch and oak. An altitudinal migrant of the north-west Himalayas east through Kashmir and Spiti to Garhwal.

ASIAN BROWN FLYCATCHER *Muscicapa dauurica* 14cm

An ashy-brown flycatcher with a strikingly large eye and conspicuous white eye-ring. Dirty white below, tinged with grey on breast and flanks, faintly streaked with ashy-brown. The prominent white throat is a useful indication of its presence when perched in thick foliage. Usually solitary, quiet and unobtrusive, and more crepuscular than other flycatchers. Perches upright on lower branches, making aerial sallies to catch insects; flicks wings and dips tail. Fairly common summer visitor to the west Himalayan foothills, at 900–1800m, in open, mixed deciduous forest, teak plantations, forest edge, groves, overgrown nullahs and bamboo forest.

RED-BREASTED FLYCATCHER *Ficedula parva* 13cm

Male is dull brown above, whitish below, with a rufous-orange chin and throat. Female has white throat and pale buff breast. White in tail conspicuous in flight or when tail is cocked. Calls often, with sharp clicking sounds, normally a double *tick-tick*. Seen singly or in loose pairs in shaded areas. May descend to ground, but prefers low and middle branches; flicks wings and cocks tail. Launches short aerial sallies to feed on tiny flying insects; hunts until late in evening. Winter visitor along the Himalayas from Kashmir eastwards, at up to 2000m, even higher during migration.

ULTRAMARINE FLYCATCHER *Ficedula superciliaris* 10cm

Male   *Female*

Male is deep blue above, with or (eastern subspecies) without a white supercilium, and with white patch on each side of tail base, and white below with dark blue breast sides. Female mouse-grey above, with white-edged dark tail; dirty white below, with upper breast glowing white. Song a repeated *che-chi-purr*; call a soft *tick*. Usually solitary; in winter often with mixed foraging parties. Keeps to upper parts of low trees and bushes, sometimes coming to ground; tail constantly jerked. Summer visitor along the Himalayan range, to 3200m, in open forests of pine, oak and rhododendron; in winter, moves to open deciduous forests, groves, gardens and orchards.

SLATY-BLUE FLYCATCHER *Ficedula tricolor* 10cm

Tim Loseby

Slaty-blue above, greyish-white below, tinged with fulvous on flanks and breast, and with a white patch at base of tail feathers. Male has black cheeks. Call a faint *tick tick*, alarm a *chrrr*; has a three-syllable song in the Himalayas. Solitary or in pairs; seen in mixed parties during winter. Rarely ventures into open or low growth. Active, hunting in characteristic flycatcher style; feeds on insects. A bird of forests, groves and gardens, breeding in the Himalayas between 1800m and 3200m, and common in Kashmir; winters in north and central India.

VERDITER FLYCATCHER *Eumyias thalassinus* 15cm

A bold flycatcher, light sky-blue to blue-green, brighter on head and throat and darker on wings and tail, and with prominent black lores. Eyes, bill and legs are dark. The female is somewhat duller, with the chin and sides of throat spotted with white, though this is visible only with careful observation; not easily sexed without the male around. It is generally quite silent, but has a pleasing song. Prefers exposed perches, flying to a different perch after each sally. Inhabits thinly wooded areas, breeding in the mountains of the north, and wintering all over the peninsula.

SMALL NILTAVA *Niltava macgrigoriae* 12cm

A small bird, the male bright purplish-blue above and ashy below, with purple sides of head, throat and breast. The female is brown, with rufous wings and tail, and has a brilliant blue patch on each side of the neck. Has a high-pitched call. Frequently seen with mixed foraging parties of small insectivorous birds. Perches very upright. Hunts flies among trees and low undergrowth, making short sallies and catching insects on the wing in typical flycatcher manner. Found in forests with dense undergrowth throughout the Himalayan range, at up to 2100m.

RUFOUS-BELLIED NILTAVA *Niltava sundara* 15cm

A brilliantly coloured bird. Male has a deep purple-blue back and throat, a dark blue mask, black forehead and brilliant blue crown; shoulders, rump and underbody chestnut-rufous. Female is an overall olive-brown, with rufescent tail; white on lower throat diagnostic. Both sexes have blue patches on sides of neck. Call a squeaky, churring note. Mostly solitary, keeping to undergrowth; seldom seen. Often flicks wings like redstart, and bobs body. Feeds on insects. Found in dense forest undergrowth and bushes along the Himalayas, between 1500m and 3200m; winters in the foothills and adjoining plains.

GREY-HEADED CANARY-FLYCATCHER
Culicicapa ceylonensis 9cm

A lively, cheery little flycatcher with bright yellow rump and underparts. Head, neck, throat and breast ashy-grey, darker on crown, and back yellowish-green; wings and tail brown, edged with yellow. The song is an enthusiastic and spirited trill, quite loud for so small a bird. Seen singly or in pairs; also associates with the mixed-species foraging flocks of the forest. Keeps to the shady middle levels or understorey of the forest. Makes sallies after flying insects, catching them with a loud snap, and tending to return to the same perch. Inhabits wooded country and hills, at up to 2700m.

SIBERIAN RUBYTHROAT *Luscinia calliope* 15cm

A wary bird, difficult to observe. The male is olive-brown above, with a conspicuous white supercilium and malar stripe, scarlet-orange chin and throat fringed with black, a pale buff breast and whitish belly. Female like male but duller, and lacks the scarlet-orange throat patch. Hops on ground or makes short dashes; flies up into bushes. Feeds on insects and molluscs. An uncommon winter visitor, found in grassy areas and bushes in the vicinity of water, from western Nepal to the north-eastern Himalayas, at up to 1500m.

WHITE-TAILED RUBYTHROAT *Luscinia pectoralis* 15cm

Male brownish-slaty, with white forehead and supercilium, and blackish-brown tail with white base and tips; chin and throat scarlet, throat sides and breast black, rest of underparts white. Female grey-brown above, with short supercilium and whitish eye-ring, dark tail tipped white; breast smoky-grey, becoming whitish lower down. Solitary unless breeding, when male sings from exposed perch, shifting body excitedly, flying from one bushtop to another. In juniper and rhododendron scrub, especially Tibetan furze, often near streams. Breeds in Tibet and south to easternmost Ladakh, north Bhutan and north Arunachal Pradesh, at 3900–4500m; winters in the foothills.

BLUETHROAT *Luscinia svecica* 15cm

A handsome bird, but unobtrusive and shy. The male has brown upperparts and a prominent white supercilium; throat blue, with a chestnut (occasionally white) spot in the centre and a black band below, and with rufous on lower breast and at base of outer tail; buff belly. (Plumage slightly variable depending on subspecies.) Female lacks blue; has blackish malar stripe continuing into broken gorget of brown spots across breast. Call a *churr* or *chuck*. A great skulker, and extremely wary. Usually solitary, entering cover with tail almost folded over back when disturbed. Prefers damp ground and heavy cover near water, at up to 3500m; migratory.

GOLDEN BUSH-ROBIN *Tarsiger chrysaeus* 14cm

Rupin Dang

A typical robin. Male has olive-brown crown, mantle and wings, yellow supercilium, and a black band from lores through eyes and cheeks; scapulars, sides of back and rump orange; tail orange, with black central feathers and terminal band; orange below with narrow dusky markings. Female olive, with faint yellowish-olive supercilium and buff eye-ring; ochre-yellow below. Indicates its presence by croaking alarm. Usually solitary or in pairs. Hops about quietly under thickets, jerking tail and drooping wings. Very secretive, but in open ground male often perches on rocks or bushtops. Found in open conifer forests of the Himalayan range.

INDIAN BLUE ROBIN *Luscinia brunnea* 15cm

Male is deep slaty-blue above, with a white supercilium and blackish lores and cheeks, and rich chestnut on throat, breast and flanks, with white belly centre and undertail. Female is brown above, with white throat and belly and buff-rufous breast and flanks. Has a high-pitched *churr* and harsh *tack*; breeding male gives a trilling song, often from exposed perch. A solitary bird, rarely observed in pairs, and a great skulker, difficult to observe. Moves about dense undergrowth in search of insects. Hops on ground, continually jerks and flicks tail and wings. Breeds in the Himalayas, between 1500m and 3500m.

WHITE-RUMPED SHAMA *Copsychus malabaricus* 25cm

A famed songster. Male has glossy black head and back, with distinctive white rump and sides of graduated tail; orange-rufous below, with black throat and breast. Female is grey where male is black; has slightly shorter tail, and duller rufous below breast. Melodious three or four whistling notes very characteristic; variety of calls, including harsh notes. Usually in pairs. Arboreal, keeping to shaded areas and foliage, only occasionally emerging. Makes short sallies; active until late evening. Feeds on insects, and rarely nectar. Inhabits forests, bamboo thickets and hill-station gardens in the Himalayan foot-hills and terai, east of Kumaon to north-eastern India.

HODGSON'S REDSTART *Phoenicurus hodgsoni* 15cm

Male is black from forehead to ear-coverts, throat and breast. Forecrown whitish, turning ashy-grey on crown, nape and mantle; wings brown with conspicuous white patch; rump and tail rufous, central rectrices dark brown; rufous below breast. Female grey-brown above, with no wing patch; rump and tail as male; throat and breast pale grey-brown, turning whitish on belly. Widely separated pairs in winter, near streams and bushes on forest outskirts; in summer often in tall poplars, catching insects like a flycatcher. Dry, barren uplands as well as open cultivation, light forest and valley floors in the lower Himalayas; in winter, dry riverbeds in forests or cultivation.

GÜLDENSTÄDT'S REDSTART *Phoenicurus erythrogaster* 16cm

A shy, suspicious bird, particularly in summer. Male boldly coloured: pure white cap and nape, with black from forehead and face to lower breast, back and wings, the last with a white patch; belly, vent, rump and tail bright rufous. Female pale brown above, with rufous lower rump and rufous-brown tail; whitish eye-ring; pale rufous-brown below, with whitish belly. Has a short, clear song, rarely recorded. Feeds mainly on flying insects and beetles, in winter changing over to the ripe fruits of *Hippophae*. Found on barren, dry slopes, meadows and riverbeds above 5000m; male especially known to be very hardy, remaining on territory even in severe weather conditions.

BLUE-FRONTED REDSTART *Phoenicurus frontalis* 16cm

Tim Loseby

Male has bright blue supercilium and forehead, darker on crown and back, with rump and tail rufous, tail with central feathers and broad terminal band blackish. Dark blue throat and breast, orange-chestnut below. Female dark olive-brown, with pale buff eye-ring; throat and breast olive-brown, orange-brown below. Call *tik* or *prit*. Usually solitary, in loose parties after breeding. Drops to ground from rock or bush in search of insects, seeds and berries, or hops in bushes; also catches insects in air. Pumps tail. Altitudinal migrant of scrub- covered slopes of the Himalayas at up to 5300m; in winter, prefers cultivation, clearings, gardens, scrub pasture and open forest.

WHITE-CAPPED WATER-REDSTART
Chaimarrornis leucocephalus 19cm

A distinctive species, with pure white cap contrasting with the surrounding black plumage, bright chestnut rump and underparts from breast downwards, and tail also chestnut but with a black band terminally. Sexes alike. Has a shrill whistle. Seen singly or in pairs beside hill streams which are exposed, and not covered by forest. Highly territorial. Catches insects carried by the current and those flying above, making rapid sallies. A bird of rocky streams in the Himalayas, at up to 5000m; in winter descends to the Himalayan foothills.

PLUMBEOUS WATER-REDSTART *Rhyacornis fuliginosa* 12cm

Male *Female*

A small bird of torrential hill streams. But for the chestnut tail and rufous lower belly, the adult male is all dull slaty. The female is grey-brown above, white and mottled slate below, with two rows of white spots on the wing, and a pale eye-ring; tail white with a terminal brown triangle. Has a *kreee* threat call, also a sharp *ziet-ziet* call. Found individually or in pairs along torrents and rushing streams. Very territorial. Pursues insects over stones and boulders, or in the air, catching them in a short sally, even from surface of water. Crepuscular, and active even well after dusk. Occurs in the wet zone of the Himalayas.

GRANDALA *Grandala coelicolor* 23cm

Male's bright purple-blue plumage with black lores, wings and square tail not assumed until second year. Female, immature and subadult brown, heavily streaked white from head to breast; white wing patch best visible in its light, almost swallow-like flight. Generally quiet, with call *tji-u*; more vocal in breeding season, when same call or variants repeated in series, e.g. *galeb-che-chew-de-dew*. Shy and wary in summer, singly or in pairs, and feeds mainly on insects and larvae; bolder in winter, in large flocks of up to several hundred, and diet entirely *Hippophae* berries. High-altitude bird of damp, rocky hill slopes or meadows, at up to 5500m and above; in winter descends to lower valleys, though not below 3000m. Ranges from central and eastern China to Ladakh and Arunachal Pradesh.

LITTLE FORKTAIL *Enicurus scouleri* 12cm

A bird of rocky mountain streams, waterfalls and small shaded forest puddles. Black above, with white forehead, white band in wings extending across lower back, white uppertail-coverts, and white outer feathers to slightly forked short tail; black throat, white below. Prominent white legs. Rather silent save for a rarely uttered sharp *tzittzit* call. Seen singly or in pairs. Moves energetically on moss-covered and wet, slippery rocks, constantly wagging and flicking tail. Occasionally launches short sallies; also plunges underwater, dipper style. Feeds on aquatic insects. Breeds throughout the Himalayas, at 1200– 3700m; descends to about 300m in winter.

BLACK-BACKED FORKTAIL *Enicurus immaculatus* 23cm

Balram Thapa

A long-tailed, bulbul-sized bird. Head and upperparts black, with white forehead and supercilium, white rump extending as a white band across black wings, white tips to secondaries; long tail deeply forked, outer feathers white, inner ones black with white tips creating graduated effect; throat black, rest of underparts white. Has a sharp *kurt-see* call; often gives brief song. A solitary bird, seen hopping about on stones and boulders, constantly wagging its tail. Feeds on insects. Found along wooded streams in the Himalayan foothills from Garhwal eastwards, at up to 1400m.

SPOTTED FORKTAIL *Enicurus maculatus* 25cm

Easily separated from similar-sized forktails by white-spotted black back. Otherwise black, apart from white forehead and forecrown, broad white wingbar and rump, and white belly and vent. Deeply forked, graduated black and white tail. Solitary or in scattered pairs. Active, moving on boulders at water's edge or in mid-stream, long forked tail swayed but mostly kept horizontal; rests in shade of damp forest undergrowth. Flies low over streams, calling shrill, screechy *kree* (also squeaky notes when perched). Feeds on aquatic insects and molluscs. Breeds on boulder-strewn forest rivers in the Himalayas, mainly at 1200–3600m; descends to about 600m in winter.

GREY BUSHCHAT *Saxicola ferreus* 15cm

Male   *Female*

A typical bushchat. Male has a pied appearance: dark ashy-grey mixed with black streaks above, and white below, suffused with grey on breast and flanks; sides of head black, with white supercilium, white patch on wing separating the black quills from rest of upperparts, and black tail margined with white. Female has the grey and black replaced with lighter and darker brown, with rusty uppertail-coverts and outer tail; underparts lightly suffused with brown, darker on breast. Call a short repeated *zeee-chunk*. Feeds mainly on insects and seeds. A bird of scrub-covered and open hillsides along the Himalayan range, at up to 3400m.

VARIABLE WHEATEAR *Oenanthe picata* 17cm

A polymorphic species, with various colour phases. Male has white rump and tail-coverts, and white tail except for black central feathers and broad terminal band; rest of plumage either all black, or with white belly, or with white belly and crown. Female is sooty-black, grey-brown or earth-brown; belly is usually buffish. Good mimic: song often an assortment of imitations; pleasant trilling in winter. Mostly solitary. Bobs body when perched; flies to ground to pick up insects. Aggressive, chasing others from its feeding sites. Mostly a winter visitor, in dry, barren open country and near villages in north-west India and Pakistan; breeds in north Baluchistan, Chitral and Gilgit.

DESERT WHEATEAR *Oenanthe deserti* 15cm

Inconspicuous from afar. Head, back and underparts pale sandy, with white supercilium contrasting with black lower face, chin and throat; wings blackish-brown; rump to tail base sandy-whitish, rest of tail black. Female paler, lacks black on face, chin and throat. In breeding season, male sings repeated mournful *teee-ti-ti-ti*; boldly approaches intruders at nest, uttering *chuck-churrr* alarm. Often pumps tail when perched, then suddenly flies upwards to catch prey, or descends to ground to feed on beetles and insects. Prefers rocky slopes and sandy plains with scattered *Caragana* bushes. Summer visitor to the north-western Himalayan plains, to 5000m and above.

CHESTNUT-TAILED STARLING *Sturnus malabaricus* 21cm

A small sociable member of the starling group. Head pale grey; upperparts darker grey with a brownish tinge; underparts rusty-red, with throat and breast reddish-grey. Beak dark with yellow tip. Western Ghats race *blythii* has the head and breast white. Arboreal, descending to feed on fruits, insects and flower nectar in bushes, and occasionally to ground. Like many other birds, attracted to emerging termites, and to trees in bloom that provide nectar. Gathers in large flocks, sometimes with other species of the family. Subject to seasonal movements throughout the Indian peninsula, at up to 2000m.

JUNGLE MYNA *Acridotheres fuscus* 23cm

Very like the Common Myna (*Acridotheres tristis*), with large white wing patch, though overall greyer-brown. Prominent tuft of black feathers on forehead; broadly white-tipped tail and yellow iris are diagnostic. Call not much different from Common Myna's. Occurs in family parties and flocks of about 10–30, often with grazing cattle. Forms large communal roosts in sugarcane fields and reedbeds, shared with noisy gatherings of Common Mynas and other birds. Partial to nectar, and is an important cross-pollinator for many plant species; forehead tuft acts as effective pollen brush. Also takes fruits, berries, grain, insects. Found throughout the lower Himalayas.

COMMON HILL-MYNA *Gracula religiosa* 29cm

Gen. R.K. Gaur

A famed talker, but a noisy bird of the forest. Larger than other mynas. Jet-black, with bright orange-yellow beak, pale yellow legs, yellow facial skin and fleshy wattles on nape and face sides, and a white wingbar. In small flocks or pairs, occasionally gathering in large numbers on fruit-laden trees. Keeps mostly to the canopy; very rarely seen on shrubbery, and hardly ever comes to the ground. Towards sunset, moves to the tall trees that tower over the forest canopy, calling and screeching, before flying to roost sites. Inhabits moist deciduous and evergreen forests, at up to 2000m.

CHESTNUT-BELLIED NUTHATCH *Sitta cinnamoventris* 12cm

A bluish-slate and chestnut bird which runs mouse-like on trunk and branches of trees, head up or head down, at times even upside-down on the underside of a branch. Prominent white patch on cheek, and a black line extending from upper mandible to nape. Has squeaky whistling and chirpy calls, and a whistle-like song. Very active, hunting for insects on the bark of trees, even in places seldom examined by other species. Often joins mixed foraging flocks, or associates with other individuals of its own species in small numbers. Found in deciduous and open forests, and other woodland habitats, throughout the Himalayas, at up to 1000m.

WHITE-TAILED NUTHATCH *Sitta himalayensis* 13cm

A bluish-slaty bird with black eye-stripe and a distinct white patch at base of tail, spread tail appearing black with white spots. Throat buffish, becoming ochraceous on breast and deep rufous below. White patch on underwing visible in flight. Weak *chip chip* call; in breeding season a clear, tit-like whistling song. In pairs or scattered parties, often with mixed bands of tits and small babblers. Seen mostly on moss-covered branches, moving jerkily along upper- or underside, prying into crevices or under moss for insects; also eats nut kernels and seeds. Found in deciduous or evergreen forest, preferably mossy, in the Himalayas from Chamba eastwards, to 3400m.

WALLCREEPER *Tichodroma muraria* 16cm

An inconspicuous ashy-coloured bird with short, dark, white-framed tail, blackish-brown wings, and black chin, throat and upper breast fading to dark grey underparts. Its beauty is seen when it opens its broad, rather long, rounded wings to reveal bright crimson shoulders and flight-feather bases and striking white primary spots. Call a feeble *ti-o-u-ti-o-u*; song *zee-zee-zee-tui*. Spurts in jerky zigzags along cliff faces, its long, thin, curved beak perfectly adapted for picking spiders, grubs and insects from narrow fissures and cracks. Resident of steep cliffs along narrow gorges and mountain rivers throughout the Himalayas, at 3300m and above, descending to foothills in winter.

BAR-TAILED TREECREEPER *Certhia himalayana* 12cm

A small forest bird that spends almost its entire life on tree trunks. Streaked blackish-brown, fulvous and grey above; pale supercilium and broad wingbar; white chin and throat, dull ash-brown below. Best recognized by dark brown barring on pointed tail. Gives long drawn-out squeak, somewhat ventriloquial; loud but short song, one of earliest, heard long before other species have begun to sing. Solitary or several in mixed parties. From near base of tree usually climbs to mid-height, then moves to another tree, checking crevices and under moss for insects and spiders; may examine moss-covered rocks and walls. In temperate forests of the Himalayas, east to west Nepal, from about 1600m to timberline; descends in winter.

EURASIAN WREN *Troglodytes troglodytes* 9cm

Göran Ekström

A tiny, closely barred, skulking bird. Short, erect tail distinctive. Brown above; paler below, whiter on belly. Quite noisy: fairly loud *zirrr-tzt-tzzt* alarm notes; shrill rambling song, sometimes uttered also in winter, even in snow. Usually solitary. Very active, but also very secretive; hops jerkily on boulders or moves mouse-like among dense bushes, taking cover if approached closely. Feeds on insects. A Himalayan bird, found in thickets, dense cover, mossy growth, rocky ground and vicinity of mountain habitation, at 2700–3900m; sometimes visits gardens; descends to about 1200m in winter, especially in the western Himalayas where it can be seen on the lower rocky slopes.

FIRE-CAPPED TIT *Cephalopyrus flammiceps* 9cm

Joanna Van Gruisen; Fotomedia

A tiny, highly active bird. Male olive-yellow above, with scarlet-orange cap, chin and throat and two yellow wingbars; yellow throat and breast, whiter-buff below. In winter, no scarlet on crown and duller yellow below. Female has yellowish rump, wingbars, and edges of outer tail feathers. Has faint twittering song. In small parties, sometimes with other species. Extremely active, flits about in canopy foliage, clings upside-down; behaviour very like leaf warbler. Feeds on small insects, buds. Breeds on forested hillsides and orchards in the western Himalayas, from Kashmir to Garhwal, at 1800–3700m; winters in central India.

RUFOUS-NAPED TIT *Parus rufonuchalis* 13cm

One of the commonest tits of higher regions. Has black crown, crest and sides of neck, white cheeks, and white nuchal patch tinged rufous near mantle; rest of upperparts grey, with no wingbars; throat, breast and upper belly black, lower belly grey, undertail-coverts and patch on flanks rufous. Loud, cheerful *gypsie-bee gypsie-bee* call or mellow *pipit-snippit*. Common in mixed foraging parties. Takes berries and seeds to regular spots on branches to hammer out the kernels. Lives in fir, pine and mixed oak and spruce forests in the west Himalayas, from north Baluchistan through Kashmir to Garhwal and Kumaon.

SPOT-WINGED TIT *Parus melanolophus* 11cm

A distinctive tit. Has black crest and sides of neck, white cheeks and nape patch, dark grey back, rusty-white double wing-bar (spots), black throat and upper breast, slaty belly, and rufous flanks and undertail. Call a two-note *te-tni* or faint *tzee-tzee*; also a whistling song. Usually part of mixed foraging bands of small birds. Restless, always on the move. Hunts in canopy for insects and berries; also descends to ground. Found in coniferous forests in the west Himalayas, east to west Nepal, from 2000m to 3800m; descends in winter to foothills, where it frequents mixed forests.

GREY TIT *Parus cinereus* 13cm

A grey and white bird with black head, neck and throat extending in a rather broad band down dirty white underparts to its vent, prominent white patch on cheek and faint patch on nape. Back grey, wings dark grey-brown crossed by white bar, and tail dark grey with white outer feathers. Its cheerful *whee-chi-chi, whee-chi-chi* call is audible throughout its forest habitat. Searches foliage and bark of trees methodically for insects, caterpillars and seeds, hanging acrobatically from twigs. Various subspecies are widely distributed over the Indian subcontinent; in the Himalayas, it is found at up to 3700m.

GREEN-BACKED TIT *Parus monticolus* 13cm

A typical tit, with white, greenish-yellow, slaty-blue and black in the plumage. Underparts mostly yellow, but pattern similar to that of the Great Tit: head and median ventral band black, and prominent white cheek patch; rump, wings and tail slaty-blue, the last two also with some white. Calls lively and variable, not different from those of tits in general. A bird of open forests, scrub with trees and suchlike habitats, but generally preferring more moist conditions than the Great Tit; found at up to 3900m all along the southern slope of the Himalayan range.

YELLOW-CHEEKED TIT *Parus spilonotus* 14cm

An active, arboreal bird. Olive-green back, black crest (faintly tipped with yellow), black stripe behind eye, and broad black central band from chin to vent; bright yellow supercilium, nape patch and sides of underbody. Seen in pairs or small flocks, often with other small birds, feeding in the foliage on insects and berries. Sometimes enters gardens. Utters cheerful, musical call notes. Found in forests of various types from the western Himalayas to east Nepal, between 1200m and 2500m; descends in winter.

SULTAN TIT *Melanochlora sultanea* 20cm

Simon Harrap

A striking bird. Male black above, with yellow crown and crest (erect when excited); black throat and upper breast, yellow below. Female has deep olive wash to upperparts and throat, latter also with some yellow. Noisy: loud, whistling *cheerie-cheerie* or other shrill notes, often mixed with harsh *churr* or *chrrchuk*; varied chattering notes. Small bands, often in mixed foraging flocks. Active and inquisitive canopy feeder: clings sideways and upside-down, checks foliage and dark crevices for insects, berries and seeds; descends to tall bushes. Found in mixed evergreen forests in the Himalayan foothills, from east Nepal to north-east India, to about 1200m, sometimes to 2000m.

BLACK-THROATED TIT *Aegithalos concinnus* 10cm

A distinctive bird. Chestnut-red crown and nape, with characteristic broad white supercilium, and black-and-white face sides and throat; back grey, wings and tail brown, tail with white tip and outer feathers; buffish yellow-red below throat. Has incessant faint *trr...trrr* and *check-check* calls. Highly sociable, in small groups, almost always part of mixed foraging flocks, though sometimes keep to themselves; often rather tame and confiding. Fidgety, overactive, ever curious, checking leaves, branches and crevices for insects and small fruits. Found in open forests, gardens and secondary growth in the Himalayas, at up to 3000m, to 3500m in its eastern limits.

COMMON SAND-MARTIN *Riparia riparia* 13cm

A modest-looking bird, grey-brown above, becoming darker towards wings and tail, and white below, with a broad grey-brown band across breast. Call a hard *brret*, rather harsh in tone, usually given on the wing around nest colony; twittering song. Gregarious, almost always seen in flocks, which often perch on telegraph wires; individual birds occasionally stray long distances. Feeds on small insects captured in flight. Occurs by sandy cliffsides, sand banks along watercourses. Patchily but widely distributed in the northern/central Indian subcontinent, at up to 4500m in summer.

PLAIN SAND-MARTIN *Riparia paludicola* 11cm

Göran Ekström

Very like the Common Sand-martin in all respects. Dirty white below, the broad, grey-brown breast band (see Common) absent or virtually so. Has a similar harsh *brret* call, and also a twittering song, usually in flight around the breeding areas. Gregarious, in flocks, which frequently perch on wires. Can be seen flying around sand banks along watercourses. Individual birds rarely stray far. Feeds on insects captured in flight. Found in the vicinity of water, in the north and north-west outer Himalayas, at up to 4200m.

EURASIAN CRAG-MARTIN *Ptyonoprogne rupestris* 14cm

Uniform ashy-brown upperparts from head to square-cut tail. Chin and part of throat are streaked whitish-brown; remaining underparts pale brown. Distinct white spots near tips of uppertail feathers (when these spread) allow easy identification in flight. Immatures more rufous. Call a soft *chip-chip-chip*, uttered mainly in flight. Forages in swallow-like flight, feeding only on flying insects. Summer visitor around cliffs and steep rocky valleys, preferably next to watercourses, throughout the Himalayan range, at up to 5000m.

BARN SWALLOW *Hirundo rustica* 18cm

Sparrow-sized with a long forked tail. Glossy steel-blue above, pale pinky-white below; forehead to throat chestnut, bordered below with a blue-black breast band; white tail spots in flight. Flight swift, a few rapid beats followed by graceful glide. Pleasant twittering call notes uttered on wing or from perch. Highly gregarious in winter, often in large, close-packed swarms perched along wires. Catches insects in low flight over meadows or near water surfaces; also feeds on ants on ground. In open country near rivers and jheels, also in cultivation and human habitation. Breeds in the western Himalayas, east to Nepal; it is common around West Himalayan hill stations and upland villages.

WIRE-TAILED SWALLOW *Hirundo smithii* 14cm

One of our prettier swallows. Pure white of underparts contrasts strongly with glistening deep purplish-blue of upperparts; rufous cap from lores to nape. Shafts of the two outer tail feathers project a good 10cm beyond rest of tail (projection shorter on female and absent when in moult). Juvenile brown, showing blue in patches, and with minimal tail-streamers. Habits and habitats as other swallows, but prefers vicinity of water; roosts with wagtails and other swallows in reeds growing in water. Resident in the Indian subcontinent, at altitudes of up to 3000m.

GOLDCREST *Regulus regulus* 8cm

Göran Ekström

A tiny bird of conifer forest. Male greyish-olive above, with prominent black-bordered golden-yellow median stripe on crown and broad whitish eye-ring; yellow in wings and tail, and two yellow-white wingbars; whitish below. Female's crown-stripe yellow. High-pitched squeaking *tsi tsi tsi* call diagnostic. In pairs or small flocks, often with other small birds, in conifer canopy; also hunts in low branches and tall growth. Restless, moving energetically, occasionally hovering, in search of insects. Himalayan coniferous forests, at 2400–4000m; considerable altitudinal movement, descending to 1500m, at times even 1200m, in winter, when also in orchards.

BLACK-CRESTED BULBUL *Pycnonotus flaviventris* 19cm

An arboreal bird which has a glossy black head, throat and crest, olive-yellow nape and upperparts, and yellow underparts. Tail mostly brown. Pale yellowish-white eye a useful feature when bird is seen at close quarters. Utters cheerful whistles; song is a *weet-tre-trippy-weet*. Usually seen in pairs or small parties, sometimes with other birds, hunting for insects, especially flying ants, or feeding on fruit. Found in forests, clearings and orchards in the Himalayas, from Himachal Pradesh eastwards, at up to about 2000m, exceptionally to 2400m.

Rupin Dang

RED-WHISKERED BULBUL *Pycnonotus jocosus* 20cm

Along with Red-vented (*Pycnonotus cafer*), the most common bulbul of the region. Brown above and white below, with pointed black crest, small red patches on cheeks (whiskers) and red vent; dark patches on breast sides (Western Ghats race has almost unbroken pectoral band). Lively and energetic, enlivening its surroundings with its cheerful liquid (almost yodelling) notes; its more musical calls are readily distinguishable from Red-vented's. Tame and confiding in some areas. Takes fruit, nectar and insects. Found in gardens, jungles and parks from Garhwal east along the Himalayan foothills, to about 1500m; widespread in the Indian subcontinent.

HIMALAYAN BULBUL *Pycnonotus leucogenys* 20cm

Essentially a north Indian species, this bulbul, like the others, adds much cheer to its surroundings by its lively calls. Brown upperparts and crest (varying in shape and length in different regions), and white underparts; white cheeks, and yellow vent. Seen in pairs or small parties. Can become quite tame and trusting in some areas. Found among scrub and light forest, in drier habitats than most of the other bulbuls. Takes berries and other fruits, insects and flower nectar. A common bird all along the southern slopes of the Himalayan range, at up to 2400m.

MOUNTAIN BULBUL *Hypsipetes mcclellandii* 23cm

Frederik Smetacek

A myna-sized bird with a dark brown crown with fine whitish streaks, and slightly crested. Rest of upperparts olive-green; throat grey, heavily streaked with white, breast cinnamon with fine white streaks, belly paler, and undertail-coverts yellowish. Has a pleasant call of varied notes. Less sociable than other bulbuls, usually in pairs or small parties, often in mixed foraging flocks. Keeps mostly to higher branches, but descends to fruiting trees in search of berries. Found in the Himalayas from Uttar Pradesh eastwards, at up to 2700m.

HIMALAYAN BLACK BULBUL *Hypsipetes leucocephalus* 23cm

A large bulbul, blue-grey overall, with a prominent dull black head and crest encircling the whitish-grey ear patch, a distinctly forked tail, and red beak and legs. Quite vocal, with an assortment of harsh, screeching and whistling calls. Gregarious, moving about in small flocks, which can grow into large aggregations on occasion. Generally keeps to the canopy and upper layers of the vegetation. Takes fruit, flower nectar and insects, the last sometimes in the manner of a flycatcher. Inhabits hill-forest and scrub-jungle, at up to 3000m.

STRIATED PRINIA *Prinia criniger* 17cm

Brown above, and distinctly streaked. Very long, strongly graduated tail, each feather with buff tip and dusky subterminal spot (narrow bars visible only at close range). Pale fulvous below, in winter with dusky spots on sides of throat and breast; flanks olive-brown. Song exuberant but grating, almost like knife being sharpened. Usually solitary or in pairs, skulking in low bushes on grassy slopes, stony ravines, terraces and open pine forests. Feeble jerky flight. When disturbed, rarely flies far before falling headlong into another bush, or down steep hillside, tail doubled over back, wings pulled in. Feeds on insects. Common resident, subject to vertical movement in the Himalayas, from the north-west east to Assam, at up to 3000m.

ORIENTAL WHITE-EYE *Zosterops palpebrosus* 10cm

A small nectar-feeding bird with a soft call. Olive-green upperparts, fading into yellowish-green head and throat, with a prominent white ring around eye; grey-white from belly to vent. Song starts inaudibly, grows louder, then fades away again. Forms small flocks which move from tree to tree in almost 'follow-my-leader' fashion. Essentially arboreal, feeding on flower nectar, insects, and the soft pulp of fruits opened up by other animals; comes down to large shrubs when attracted by flower nectar. Builds cup nest in forks of twigs. Found in forests, gardens, groves and secondary growth, at up to 1800m.

CHESTNUT-HEADED TESIA *Tesia castaneocoronata* 8cm

A tiny wren-like bird, one of the smallest in Indian subcontinent, with extremely short tail. Olive-green above, with bright chestnut forehead to nape; throat bright lemon-yellow, breast and belly olive-washed yellow, flanks olive. Chattering *chiruk-chiruk* or single loud, piercing *tzeet*, repeated when alarmed; jerks body when calling. Solitary, in pairs during breeding season. Shy and elusive. Frequents undergrowth in high, rather open forest and dark ravines near streams; keeps near ground, hopping about in cover or at times on moss-covered boulders or logs. Feeds on insects and spiders. Resident along the Himalayas, from Kangra eastwards, to 3900m; subject to seasonal movement.

BROWNISH-FLANKED BUSH-WARBLER *Cettia fortipes* 11cm

Rufescent olive-brown above, with buffy eyebrow and dark line through eye, and dull whitish below, tinged ashy-brown on throat, with buff-brown flanks and undertail. Call *tyit-tyu-tyu*; diagnostic song, *weeee... chiwiyou*. A shy, secretive bird, sneaking in undergrowth, feeding on insects; rarely seen, its presence usually confirmed by its voice. Found in undergrowth on hillsides, open forest, forest edges, bamboo, also tea gardens along the Himalayas, breeding from 2000m up to about 3300m; moves to foothills in winter.

Göran Ekström

GREY-SIDED BUSH-WARBLER *Cettia brunnifrons* 11cm

A rufous olive-brown bird with rufous-chestnut crown, long buff supercilium and dark stripe from lores through eye. Throat and belly white, breast and sides grey, flanks and vent olive-brown. Has a soft *tsk tsk* call. Though difficult to observe, can be inquisitive at times, and may sing from perch on top of bush or rock. When disturbed, rarely rises much above ground or flies far. Feeds on insects among bushes; also creeps on ground. Found among dwarf rhododendron, bamboo and other bushes in forest clearings and margins, also open forest and tea gardens in winter. Common vertical migrant of the Himalayas, from Uttar Pradesh east to Bhutan, at up to 3700m.

BOOTED WARBLER *Hippolais caligata* 12cm

Tim Loseby

A very active bird. Dull olive-brown above, with short pale supercilium, and pale buffy-white below. Solitary or up to four birds together, sometimes in mixed bands of small birds. Harsh but low *chak chak churrr* call, heard throughout day; also a soft jingling song, sometimes heard before departure from winter grounds. Agile, moving among leaves and upper branches in search of insects. Very like a leaf warbler in behaviour, but calls diagnostic. Found in open country with acacia groves in, to 2000m and above; breeds in the north-west, winters all over the Indian subcontinent.

WHITE-BROWED TIT-WARBLER *Leptopoecile sophiae* 11cm

Tim Loseby

An active little bird. Has brown crown and nape, prominent white eyebrow, bluish face sides and purplish on the throat, breast and flanks (race *obscura* from the central Himalayas darker and more purple than nominate race in west). Fairly sharp tseet call; pleasant chirping song. Quite shy. Can hang upside down like a tit when foraging. Builds a ball-shaped nest close to the ground in bushes or undergrowth. Occurs locally at high altitude above the treeline in dwarf rhododendron or juniper. Found mainly in Ladakh and north-west Nepal, usually above 3500m, locally above 4500m; can descend to 1800m in winter.

COMMON CHIFFCHAFF *Phylloscopus collybita* 10cm

Pale olive-brown above, with a short white supercilium, and dull whitish below, washed with buff on breast and flanks. No wingbars. Has a plaintive *tweet* or *wheet* call. Found singly or in parties of up to ten birds, moving restlessly from bush to bush or hopping on ground in search of insects, flicking wings and tail in movements typical of genus; forages also in standing crops, in stubble and among waterside vegetation, including partially submerged bushes. Abundant winter visitor to the lower Himalayas, from west Pakistan eastwards to the Bhutan foothills.

TICKELL'S LEAF-WARBLER *Phylloscopus affinis* 10cm

A dark olive-brown bird with prominent long, yellow supercilium, almost canary-yellow below. No wingbars. In breeding season sings constantly from bushtops, a single note repeated rapidly five or six times. Found singly or in pairs in summer, often in loose parties in winter, feeding in low bushes close to ground on small beetles and other insects; at times clings upside-down to twigs, or makes short sallies into air. Abundant summer visitor throughout the Himalayas, at up to 4500m, frequenting willow, juniper and rhododendron forests, forest edges and *Caragana* bushes.

HUME'S WARBLER *Phylloscopus humei* 10cm

Has two wingbars, the upper one fainter. Greenish-olive above, with faint crown-stripe, striking pale yellowish supercilium and eye-ring; underparts whitish, tinged with light yellow. Has a sharp, rather long *tiss-yip* call. Forages for insects high up in treetops, also in low bushes; may be seen fluttering against bark or descending to ground. Common summer visitor to glades in mixed or coniferous or birch forests of the Himalayas, from north-west to Kashmir and Garhwal, at up to 4000m; prefers gardens, orchards and dry deciduous forest in winter.

GREENISH WARBLER *Phylloscopus trochiloides* 10cm

Toby Sinclair

A small warbler, very similar in plumage to the Large-billed Leaf-warbler. Brownish-olive above, with yellowish supercilium and single short wingbar; dull yellow below. Best identified by its call, a squeaking, fairly loud *tchiewee* or *chee...ee*. Seen singly or in scattered, mixed parties. Spends most of its time foraging for small insects in leafy upper branches of medium-sized trees. Breeds in forests and groves in the Himalayas, between 1800m and 4300m.

LARGE-BILLED LEAF-WARBLER *Phylloscopus magnirostris*
13cm

Göran Ekström

A big leaf-warbler, very similar to the Greenish Warbler. Differs in larger size, colour of lower mandible, which is brown instead of yellow, and legs, which are slate-grey instead of brown. Call most reliable distinction, a characteristic *dir-tee*, the first note clearly lower; also has a ringing five-note song in summer. Keeps to forests in vicinity of streams, usually high in the foliage. Found in summer along the Himalayas from the Afghan frontier eastwards to Bhutan, at up to 3600m; winters throughout the Indian peninsula.

BLYTH'S LEAF-WARBLER *Phylloscopus reguloides* 9cm

Has two wingbars, but only one really apparent. Light greyish-olive above, wings more yellowish, shoulders yellow; pale median crown-stripe with broad dusky-olive band on each side, blackish on nape, a conspicuous pale yellow supercilium, and dark line through eye; cheeks pale yellow; greyish-white below, breast and belly slightly streaked with yellow. Call *kee-kew-i*, repeated frequently. Singly or in pairs in winter, often with mixed flocks, searching for insects both in tree canopy and in bushes; also eats berries. Altitudinal migrant, breeding in oak, rhododendron, conifer and mixed forests from Kumaon to east Bhutan, at 2000–3500m; bushy areas in winter.

111

GOLDEN-SPECTACLED WARBLER *Seicercus burkii* 10cm

A tiny warbler of forest undergrowth. Olive-green above, with greenish or grey-green eyebrow bordered above by prominent black lateral crown-stripe, and greenish face sides; yellow eye-ring; completely yellow below. Fairly noisy, uttering sharp *chip-chip* or *cheup-cheup* notes. Forms small restless flocks, often in association with other small birds. Keeps to low bushes and lower branches of trees. Feeds on insects. Breeds in the Himalayas, between about 2000m and 3700m; winters in the foothills and the north-eastern Indian peninsula.

GREY-HOODED WARBLER *Phylloscopus xanthoschistos* 10cm

Rupin Dang

Fairly distinctive. Grey above, with prominent, long grey-white eyebrow and yellowish-olive rump and wings; completely yellow below. White in outer tail visible in flight. Quite vocal, its loud, high-pitched double call a familiar sound of Himalayan forests; pleasant trilling song. Pairs or small flocks, often with mixed foraging parties. Highly energetic; actively hunts and flits about in lower canopy and tall bushes. Feeds on insects, rarely small berries. A bird of Himalayan evergreen open forests, at 1000–3000m; altitudinal movements in winter.

WHITE-THROATED LAUGHINGTHRUSH *Garrulax albogularis*
28cm

A typical laughingthrush. Olive-brown above and ochraceous below, with fulvous forehead, a black mark in front of and below eye, and prominent white throat bordered below by an olive-brown band. Long rounded to graduated tail, with terminal part white. Call a low murmuring *teh-teh*, becoming a noisy chorus when disturbed. Highly gregarious, remaining in flocks even in breeding season. Feeds on ground, turning over leaves and twigs for insects; also in trees. Found in dense forests at medium and high altitudes in the Himalayas, at up to 3300m.

WHITE-CRESTED LAUGHINGTHRUSH *Garrulax leucolophus*
28cm

Larger than a myna and ummistakable. Olive-brown, with prominent white crest, head, throat and breast; black band extending from beak, through and below eye to well beyond it. Vocal and gregarious: found in flocks of around a dozen or more individuals feeding on insects, fallen fruit and the like on the forest floor, turning over leaves and fallen twigs, and keeping up a lively series of calls, now and then bursting into 'laughter'. The flock moves on in almost 'follow-my-leader' fashion, one after another flying and catching up with the others. Inhabits forest undergrowth, at up to 2200m

STRIATED LAUGHINGTHRUSH *Grammatoptila striata* 28cm

Tim Loseby

Rich brown plumage heavily streaked white, except on wings and rich rufous-brown tail. Darkish loose crest, streaked white towards front; heavy streaking on throat and sides of head, less marked from breast downwards. Very vocal: has clear whistling call of 6–8 notes; loud cackling laughter. Pairs or small parties, often with other birds in mixed, noisy parties. Feeds both in upper branches and in low bushes, on insects and fruits; seen to eat leaves. Shows a marked preference for certain sites in forest. Found in dense forests, scrub and wooded ravines from western Himachal Pradesh east along the entire Himalayan range, at up to 2700m.

STREAKED LAUGHINGTHRUSH *Trochalopteron lineatum* 19cm

A bulbul-sized laughingthrush, grey and chestnut with pale streaks. Ear-coverts, wings and tail bright reddish-brown and characteristic of the species; pale feather shafts prominent. Has variety of harsh squeaky calls, a *chit-chit* or *chitrr-chitrr* or *pitt-wee-er* and variants. Lives in small flocks. Usually observed low down in the vegetation, only occasionally rising higher; tends not to fly from tree to tree, preferring to move down into the undergrowth, across to next bush and then clamber up again. Prefers bush-covered slopes and open forest, rarely coming to gardens. Found along the entire Himalayan range, at up to 3900m.

VARIEGATED LAUGHINGTHRUSH *Trochalopteron variegatum* 24cm

A well-marked species. Olive-brown above, with grey, black and white head; grey, black, white and rufous in wings and tail; black chin and throat, bordered with buffy-white. Vocal: clear, musical whistling call of three or four syllables, also harsh squeaking notes. Small flocks of up to a dozen or more on steep bushy hillsides. Keeps to undergrowth, occasionally clambering into leafy branches; wary, secretive, not easily seen. Weak flight, as most laughingthrushes. Feeds on insects, fruit, rarely nectar. Found in forest undergrowth, bamboo and hill-station gardens, from western Himalayas, east to central Nepal, at 1200–3500m. Breeds at 2000–3200m.

BLACK-FACED LAUGHINGTHRUSH *Trochalopteron affine* 25cm

Diagnostic blackish face and throat with contrasting white malar patches, neck sides and eye-crescents. Rufous-brown above, finely scalloped on back; olive-golden flight feathers tipped grey; rufous-brown below throat, with scaly markings. Has various high-pitched notes, chuckles; rolling *whirrr* alarm; four-note song, rather plaintive. Pairs or small bands on ground and in low growth; also ascends into middle level of trees; noisy when disturbed. Feeds on insects, berries and seeds. Prefers forest undergrowth, also dwarf vegetation in higher regions. Found in the Himalayas, east from western Nepal, at up to 4100m; descends to about 1500m, or lower, in winter.

RUSTY-CHEEKED SCIMITAR-BABBLER
Pomatorhinus erythrogenys 25cm

Göran Ekström

An elusive bird, more often heard than seen. Olive-brown above, with orange-rufous (rusty) face sides to flanks and undertail-coverts; rest of underbody mostly pure white. Long bill, downcurved towards tip, is conspicuous. Vocal: call a mellow, fluty, two-note whistle, *cue-pe, cue-pe*, followed by single (sometimes double) note reply by female; guttural alarm and liquid contact note. Found in small flocks in thick forest and undergrowth, hopping on jungle floor; turns over leaves or digs with its strong beak; sometimes hops into leafy branches. Feeds on insects, grubs and seeds. Inhabits the Himalayan foothills, at up to 2200m or even 2600m.

PIN-STRIPED TIT-BABBLER *Macronus gularis* 11cm

Simon Harrap

A greyish-olive bird with tawny-olive cap and wings, and pale yellow lores and supercilium. Underparts pale yellow, chin, throat and breast dark-streaked, with flanks olive-buff. The noisiest of babblers, thus easy to locate; has a rich, mellow but metallic call. Parties of up to a dozen or more, often in mixed-species flocks; singly or in pairs during breeding season. Actions and behaviour tit-like. Searches for insects in high canopy or among bamboo, flocks moving quickly from tree to tree. Found in both light and dense forest among undergrowth and long grass at low elevations in the eastern Himalayas, from west Nepal bhabars, duns and terai eastwards.

SILVER-EARED MESIA *Mesia argentauris* 15cm

Brightly coloured. Black head with silver ear-coverts and yellow forehead; orangey throat fading into yellow-grey breast, with belly and flanks grey; crimson wing patch, crimson uppertail- and undertail-coverts. Female has yellowish tail-coverts. Incessant *chirrup* calls while feeding; also long-drawn *seesee-sweewee*. Flocks of up to 20 birds flit from tree to tree, actively searching for insects, seeds and berries; tit-like acrobatic behaviour, also flycatcher-like sallies. Found in scrub and open clearings in evergreen forests, from the central Himalayas east to Arunachal Pradesh, from foothills to 2000m.

RED-BILLED LEIOTHRIX *Leiothrix lutea* 13cm

An attractive little bird. Male olive-grey above, with dull buff-yellow lores and eye-ring, and yellow, orange, crimson and black in wings (crimson greatly reduced in western race); black tail forked; yellow throat and orange-yellow breast diagnostic, grey to dirty white below; scarlet beak. Female has yellow instead of crimson in wings. Lively and vocal, uttering wistful, piping *tee-tee-tee*; also sudden explosive notes; song a musical warble. Small parties, often with mixed flocks. Rummages in undergrowth, frequently moves up into leafy branches; eats insects and berries. Forests, bushy hillsides and plantations, from Kashmir to extreme north-east, at 600–2700m.

WHITE-BROWED SHRIKE-BABBLER *Pteruthius flaviscapis* 17cm

Rupin Dang

A stocky, short-tailed, fearless bird. Male ash-grey above, with black head with white postocular stripe, and black wings with chestnut inner secondaries; whitish below, lower flanks and vent vinous-brown; white inner edge of primaries in flight. Female brown-grey above and pale buff below, with grey head; outer wing yellow-green, tipped white, inner secondaries chestnut; tail green, outer feathers yellow-tipped blackish. Harsh, shrike-like churring call. In pairs when breeding, otherwise in small parties or singly in mixed groups of small insectivores. Arboreal, seeking insects among leaves, crevices and under moss in higher canopy. Inhabits broadleaf forests, in west of the region also coniferous, along the Himalayas, to 2800m.

BAR-THROATED MINLA *Chrysominla strigula* 15cm

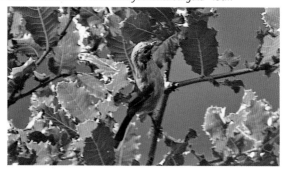

Small, active, but rather skulking bird with slightly tufted yellow-olive cap, whitish face and black malar stripe. Back grey-olive; wings with orange-yellow and black; tail chestnut, black and yellow. Dull yellow throat with thin black scales a good field mark. Utters mix of whistling squeaks; loud four-note song, with accent on second note. Small flocks frequently in mixed hunting parties, typical of Himalayas. Arboreal; hunts in canopy or middle levels for insects, fruits and nectar. In oak, fir, rhododendron and bamboo forests in the Himalayas, east of Kangra (Himachal Pradesh), at 800–3700m, breeding mostly above 1800m.

RUFOUS SIBIA *Heterophasia capistrata* 22cm

Rich rufous, with grey-brown back, black head with slight bushy crest, and bluish-grey wings. Long rufous tail tipped grey, with black subterminal band. Cheerful calls, a wide range of sharp whistles; rich song of 6–8 syllables during Himalayan winter. Small flocks, at times with other birds. Active gymnast, ever on the move. Hunts in canopy and middle levels, moving among moss-covered branches; leaps into air after winged insects; also takes nectar and berries. Himalayan forests, both temperate and broadleaf, at 1500–3500m, descending to 600m in severe winters.

WHISKERED YUHINA *Yuhina flavicollis* 13cm

Olive-brown above, with chocolate-brown crown and crest, white eye-ring and black moustache. Rufous-yellow nuchal collar (less distinct in western race); white below, streaked rufous-olive on breast sides and flanks. Quite vocal, a mix of soft twittering notes and fairly loud tit-like *chee-chi-chew* call. Flocks, almost always in association with other small birds. Active and restless, flitting about or hunting flycatcher style; moves between undergrowth and middle levels, sometimes ascending into canopy. Feeds on insects, berries, nectar. Found in Himalayan forests, from western Himachal to extreme north-east, at 800–3000m.

119

BLACK-CHINNED YUHINA *Yuhina nigrimenta* 11cm

Rupin Dang

A small yuhina with an erectile black crest with scale-like grey edgings, black lores and chin, and black-tipped red bill. Nape and sides of head grey; underparts pale fulvous. Utters constant buzzing and twittering calls. Very gregarious, active and noisy; in flocks of about 15–20 birds or in mixed parties, busily seeking insects in tree canopy, also low shrubs, clinging upside-down to twigs. Found in evergreen and secondary forests, especially overgrown clearings, in the Himalayan foothills from Garhwal eastwards, at up to 1900m.

HUME'S SHORT-TOED LARK *Calandrella acutirostris* 11cm

A dull-looking bird, grey-brown above with darker streaks, dull white below with finely brown-streaked breast. Pale supercilium from stout yellow-brown bill to nape. Legs pinkish-brown. In flight, shows dark brown tail with outer two pairs of feathers white, while rufous tint on rump may help distinguish this species from other short-toed larks. Call a sharp *trreee*, repeated monotonously as song. Highly gregarious; feeds on seeds and insects. Summer visitor to arid, stony and sandy plains with sparse grass cover in high areas of extreme north and north-west India, at up to 5000m; winters in the central and north Indian plains.

HORNED LARK *Eremophila alpestris* 20cm

A large, bold lark, pink-brown above and whitish from breast to vent. Distinctive black crown bands with black 'horns' at rear, dull yellow-white face and throat with black forehead bar crossing through eye to cheek, and black breast band (gorget). Female has crown streaked black, and cheeks and gorget duller. Call a plaintive soft *peo* or *tsie-rii*; in breeding season utters brief, modest song from low rock or mound. Feeds on ground, on seeds, grain and small insects. Resident of high-altitude plains of the entire Himalayan range, to 5500m and above, preferring barren habitats with sparse grass and bush cover; descends to lower valleys in winter.

FIRE-BREASTED FLOWERPECKER *Dicaeum ignipectus* 9cm

Nigel Redman

A tiny, restless bird. Male metallic blue-green-black above and buffy below, with scarlet breast patch and black stripe down centre of lower breast and belly. Female olive-green above, yellowish on rump, and bright buff below, with flanks tinged olive. Has a clicking song. Mostly solitary; may be encountered in mixed foraging flocks of small birds. Arboreal and active, flitting about in foliage canopy, visiting clusters of *Loranthus*. Feeds on berries, nectar, spiders and small insects. Breeds in Himalayan forests and orchards, from Kashmir to extreme east, at 1400–3000m; winters as low as 300m.

121

RUBY-CHEEKED SUNBIRD *Chalcoparia singalensis* 11cm

Simon Harrap

Male metallic green above, pale yellow below, with deep coppery-red ear-coverts and rufous-buff throat and breast. Female olive-green above (not metallic); paler below; usually some yellow in wings. Young similar to female, but very pale rufous (sometimes absent) on throat. Fairly loud chirping call, often given in short leaping flight. Pairs or small loose flocks; sometimes associates with other birds. Hunts restlessly among leaves and branches for insects or visits flowers for nectar. Found in evergreen forests and thick bushes in lower foothills of the eastern Himalayas, from Nepal to the north-east, at about 700–900m, rarely over 1000m.

GREEN-TAILED SUNBIRD *Aethopyga nipalensis* 14cm

A typical sunbird. Male has metallic blue-green crown and nape, bordered by maroon or crimson-brown band on neck sides and lower mantle; ear-coverts black; back and wings olive-green, rump bright yellow, with metallic blue-green tail; throat metallic blue-green, rest of underparts bright yellow, with red-streaked breast. Female olive-green, with outer tail tipped with white. Call a sharp *zig-zig*. Feeds on nectar. Resident of dense oak and rhododendron forests from the western Himalayas eastwards, breeding mostly above 1800m.

CRIMSON SUNBIRD *Aethopyga siparaja* 15cm

A beautiful sunbird. Longer-tailed male has metallic green crown and tail, deep crimson back and neck sides, and yellow rump (only partly visible); bright scarlet chin to breast striking; olive-yellow belly. Female has olive plumage, yellower below. Has sharp, clicking call notes; pleasant chirping song from breeding male (June-August). Solitary or in pairs. Active and acrobatic, hanging upside-down and sideways while probing flowers for nectar; also hovers. Moves a lot between tall bushes and canopy. Also feeds on insects and spiders. Found in forests of the Himalayan foothills, from Himachal east to Sikkim, at up to 1700m.

STREAKED SPIDERHUNTER *Arachnothera magna* 18cm

Both sexes olive-yellow, profusely streaked, with very long, curved bill. High-pitched *chee-chee-chee* call; loud *which-which...* song, sounding somewhat like a tailorbird (*Orthotomus*). Usually solitary; sometimes twos or threes in vicinity. Active, moving considerably between bush and canopy. Extraordinary bill is specially adapted to its nectar diet: wild banana blossoms are a favourite, the bird clinging upside-down on the bracts; also feeds on insects and spiders. Resident of forests, secondary growth and nullahs in the foothills from central Nepal east to the eastern Himalayas, and the north-east.

Rupin Dang

RUSSET SPARROW *Passer rutilans* 14cm

Male is rufous-chestnut above, heavily black-streaked on the back, with black chin and throat, greyish breast and flanks, and rest of the body pale yellow. Two white wingbars, one broad and one narrow. Female brown above, streaked dark brown on back, with conspicuous supercilium and wingbars; dull ashy-yellow below. Call very like that of House Sparrow (*Passer domesticus*), but less flat. Forms large flocks, which fly into trees when disturbed. Feeds on grain and seeds. Resident near sparse forests, human habitations and cultivation along the Himalayan range, at up to 4300m.

EURASIAN TREE-SPARROW *Passer montanus* 15cm

A gregarious mountain bird. Both sexes have chestnut-brown crown and nape, black patch on white ear-coverts, and black chin and throat, lacking yellow in plumage. Chirping call notes. May associate with finches; often perches on dry branches and overhead wires. Feeds mostly on ground, picking seeds and insects. Habitats include cultivation, edges of forests and mountain habitations. Found in the central and eastern Himalayas, at up to about 4000m, breeding at 1200–2600m; descends in winter.

TIBETAN SNOWFINCH *Montifringilla adamsi* 18cm

Rather a dull bird, generally grey-brown with a slightly darker-streaked back, and dark brown wings with distinct white patches on greater coverts and secondaries. Tail dark brown, the outer feathers white with broad brown band at tip. Light sandy-brown below, with throat darker, at times blackish. Legs and strong, typical finch-shaped bill dark horn-brown. Call is a straight *pink-pink*. Feeds on ground, mainly on seeds and insects. High-altitude resident of sandy or stony plateaux, boulder-strewn slopes and villages in Ladakh, Spiti and northern Sikkim, from 3800m to almost 5000m; in winter, descends to valley bottoms in large flocks.

PLAIN-BACKED SNOWFINCH *Pyrgilauda blanfordi* 15cm

A small, attractive finch with distinctive facial pattern of three black bands: one through forehead, one from lores through eyes, and one down throat, all against a whitish basic facial colour. Neck, shoulders and breast sides rufous, fading to fulvous-brown on back and rump; wings brown with some whitish; tail brown, with white sides and darker tip; underparts appear creamy-white. Has rapid twittering song, mainly during breeding season. Feeds on the ground, on seeds and insects. High-altitude resident, limited to eastern Ladakh and northern Sikkim, preferring sandy gentle hillsides, stony plains and steppes adjoining marshes and lakes above 4000m.

FOREST WAGTAIL *Dendronanthus indicus* 17cm

Olive-brown above and dull buffy-white below, with two black bands across lower throat and breast diagnostic. Wings dark brown with yellow-spotted bands; tail dark with white outer feathers. Has a *pink* or *tsif* call. Solitary or in pairs. Runs on ground, stopping with characteristic sideways movement of rear body; often observed on horizontal branches, picking insects. Food also includes small snails and worms. Found in forests, clearings and cultivation, and by streamsides. Breeds in Assam; winters in north-eastern India, to about 2000m.

ROBIN ACCENTOR *Prunella rubeculoides* 16cm

A handsome bird, dark-streaked pale brown above, with grey-brown head and throat contrasting with conspicuous rufous breast and pale cream-tinted belly and vent. Two whitish wingbars; rear flanks brown-streaked. Call a high-pitched *tzwe-e-yuu*; at start of breeding period sings a simple, chirping *si-tsi-si-tsi tsutsitsi* from bush or rock; when disturbed, takes wing with *zieh-zieh* alarm. Hops on ground, feeding on seeds and insects. Preferred habitats include damp areas in proximity of water and *Caragana* bushes. Resident at high altitudes of the Himalayan range, to 5500m; gregarious in winter, descending to valleys as low as 2000m, rarely below 1500m.

RUFOUS-BREASTED ACCENTOR *Prunella strophiata* 15cm

Overall heavily streaked dark brown, with whitish throat and belly and conspicuous unstreaked rufous breast and supercilium. Sharp trilling call, also *tszi-tszi*; short, chirping song. Flocks in winter, occasionally along with other accentors, pipits and sparrows; rather tame and confiding around high-altitude habitation. Hops on ground, flying into bushes if approached too closely. Feeds on insects and small seeds. Preferred habitats include damp grass and villages. Found throughout the high Himalayan range, at 3000–5000m; descends in winter to about 2000m, rarely below 1500m.

BROWN ACCENTOR *Prunella fulvescens* 15cm

A dull, almost sparrow-like bird with dark-streaked pale brown upperparts. Long white supercilium from lores to nape above dark brown facial mask, and two faint whitish wingbars, are only conspicuous markings; uniformly buff below, slightly ochreous on breast. Slim, sharp bill black; legs pale reddish-brown. Variable song, *tuk-tileep-tileep-tileep-tileep* or *tuk, siip, siip, siip.....tuk, tiew, tiew*, delivered from bushtop. Feeds on ground, on seeds and insects, hopping among stones and through bushes. Scarce resident at high elevations of the Himalayan range, to 5000m and above, on drier bush- and shrub-covered slopes; descends to lower valleys in winter.

MAROON-BACKED ACCENTOR *Prunella immaculata* 16cm

Tim Loseby

A small, shy, generally grey bird, with rufous olive-brown upperparts becoming maroon on lower back and rump. Forehead and crown with scaly whitish markings; tail greyish-brown; underparts ashy-brown, with lower abdomen, vent and undertail-coverts dark chestnut. Call a double-noted *zich-zit*. Gregarious in winter. Usually secretive, but restless, flying from ground to treetops. Feeds on seeds and insects. Inhabits conifer and rhododendron forests and rocky outcrops along the eastern Himalayas, between 2800m and 4200m.

FIRE-FRONTED SERIN *Serinus pusillus* 12cm

A distinctive little bird. Scarlet-orange forehead and blackish-grey crown; buffy back, streaked dark; yellow-orange rump and shoulders; yellow wing edges and whitish wingbars; sooty-brown below with grey and buff streaks, belly and flanks brown-streaked dull yellow-buff. Pleasant twittering *chrr chrr*; faint *tree, tree* call. Gregarious, and quite active; drinks and bathes often. Spends considerable time in bushes and low trees. Feeds on flowerheads, and on grass seeds on ground; also eats small berries. Found on rocky, bush-covered mountainsides in the western Himalayas, east to Garhwal, at 1500–4500m; breeds mostly at 2400–4000m.

YELLOW-BREASTED GREENFINCH *Chloris spinoides* 14cm

Gen. R.K. Gaur

A small finch, predominantly dark above and yellow below. Crown, ear-coverts and malar stripe blackish-brown, mantle and back similar but with olive tinge; forehead, supercilium, neck sides and rump yellow; tail blackish with yellow sides, and wings blackish, yellow and whitish. Female duller and more greenish. A number of lively calls include twittering *dwit-it-it* notes, also long-drawn *weeee-chu-chu*. Found in pairs or small flocks. Frequents cultivation, forest edge and hill-station gardens all along the Himalayan southern slopes, at 1100–4400m, extending up to the timberline.

EUROPEAN GOLDFINCH *Carduelis carduelis* 14cm

Slightly smaller than a sparrow, with a crimson face, grey-brown upperparts and pale underparts. Wings and tail black, wings with a large yellow patch and tail with some white, and rump white, all producing distinctive pattern in flight. Call a liquid *deedelit*; song a canary-like twitter, often given in chorus in breeding season. Keeps in pairs when breeding, and in small groups at other times. Feeds on seeds on or near ground, in orchards, on bare hillsides and in open forest, going up beyond treeline in warmer seasons. Found from the western Himalayas to central Nepal, at 2000–4000m, descending into foothills in winter.

TWITE *Carduelis flavirostris* 14cm

A dull bird, mostly fulvous-brown. Heavily streaked dark brown above, fading to pale pink rump (largely hidden by wings); wings dark brown with two buffy bars; throat plain, breast and flanks well streaked, belly more creamy. Tail forked, dark brown with whitish sides. Female lacks pink on rump. Short, stout horn-white bill. Rather loud, tinny *twite-twite* or *chew-chew*, often given in flight. Runs or hops on ground in constant search of seeds. Resident at high levels from Ladakh to northern Sikkim, to 5000m, on grassy prairies, arid stony hillsides with scattered Caragana bushes or boulder-strewn meadows; in winter flocks descend to lower areas, not below 3000m.

BRANDT'S MOUNTAIN FINCH *Leucosticte brandti* 18cm

Dark brown head, fading to paler brown upperparts (lightly streaked) and wings, latter with large whitish patch. Rump scaled rosy-pink, tail dark brown with white sides; all brownish-grey below. Typical finch-shaped bill and legs are black. Call a loud *twitt-twitt* or *twee-ti-ti*, chiefly in flight. Gregarious; feeds on ground, on seeds, shoots, some insects. Courting male swoops at female before perching next to her, erect with wings shaking or tail and wings raised high, singing melodic *pink-pink*. High-altitude resident from western Himalayas east to Sikkim, to 5400m, on stony hillsides, meadows and moraines; big flocks descend to lower valleys in winter, not below 3000m.

COMMON ROSEFINCH *Carpodacus erythrinus* 15cm

Joanna Van Gruisen

A sparrow-sized finch, male having a crimson crown and rump, more brownish neck sides, nape and wings, and crimson throat, gradually becoming whitish-buff on vent. Female like House Sparrow (*Passer domesticus*), brown above, white below, with streaked breast and flanks, and two pale wingbars. Call a bright, whistling *ti-deo, di-deo*. A typical finch, feeding on ground, flying into trees when disturbed. Joins other species, such as buntings. Takes seeds and small fruits, also nectar. Migratory, breeding in central Eurasia and Himalayas, to 4000m; found almost throughout the Indian subcontinent in winter.

BEAUTIFUL ROSEFINCH *Carpodacus pulcherrimus* 15cm

Peter Morris

A streaky brown bird with pale supercilium, much as Pink-browed Rosefinch but with streaked crown. Has single *sweet* call; also harsher notes. Usually occurs in small flocks, sometimes with other species. Feeds on the ground, on paths and clearings, taking seeds and berries; often difficult to see, and tends to 'freeze' if disturbed. Found in open forests, dwarf junipers and upland plantations in the Himalayas (less common in eastern Himalayas), breeding at about 2800–4000m; winters lower down, sometimes to 600m.

PINK-BROWED ROSEFINCH *Carpodacus rodochrous* 15cm

Male has a rosy-pink forehead and broad supercilium, a crimson-brown, unstreaked crown and eye-stripe, a streaked back, and pinkish rump and underbody. Female is streaked throughout, with a pale yellowish supercilium. Call resembles a canary's, *chew-wee*. Usually in small flocks, sometimes with other species. Feeds on seeds and berries taken on the ground, on tracks and clearings. Perches on trees and in bushes. A bird of open forests, dwarf junipers and habitation along the Himalayas, from Kashmir to eastern Nepal, at about 2800–4000m; may descend as low as 600m in winter.

WHITE-BROWED ROSEFINCH *Carpodacus thura* 18cm

Male is brown above, streaked blackish, with pink and white forehead and supercilium, dark eye-stripe, and rose-pink rump and underparts and double wingbar. Female streaked brown, with broad, whitish supercilium and single wingbar, and yellow rump; dark-streaked buffy below. Calls often when feeding, a fairly loud *pupuepipi*. Small flocks, sometimes with other finches. Forages mostly on ground, for seeds and berries, but settles on bushes and small trees. Found in treeline Himalayan forests of fir, juniper and rhododendron, breeding between 3000m and 4000m; winters at about 1800m on open bushy mountainsides.

STREAKED ROSEFINCH *Carpodacus rubicilloides* 19cm

Male   *Female*

Fairly distinctive rosefinch. Nape to back grey-brown, streaked darker, rump pinkish; wings brown; head and underparts carmine-red with distinct white spots, fading to pinkish vent. Tail brown with very narrow inconspicuous white edges. Female grey-brown above and buff below, intensely streaked all over, less towards vent; two pale wingbars. Heavy bill horn-brown. Call a loud *twink-twink*; song *tsee-tsee-soo-soo*. Feeds in pairs or small flocks, mainly on ground. Resident of higher elevations of the northern Himalayas from Ladakh to Bhutan, to 5000m and above, on arid plains and hillsides; in winter descends in large flocks to lower valleys, not below 3000m.

GREAT ROSEFINCH *Carpodacus rubicilla* 20cm

Male similar to the Streaked Rosefinch but paler, with no streaks on back or only very faint ones. Female is very pale, with white margins on outer tail feathers. Has a loud, single call note. Parties, sometimes numbering several dozen individuals, feed on the ground and in bushes, on seeds and berries. May associate with other finches. A bird of the high-altitude barren regions and scrub of the Himalayas, from Ladakh to Bhutan, breeding above 3500m and up to about 5000m; does not descend below 2000m in winter.

RED CROSSBILL *Loxia curvirostra* 14cm

Tim Loseby

A stout finch, the distinctive crossed mandibles visible at close range. Male mostly dull red, lightly marked brown above, with dark stripe through eyes, blackish wings and short, forked tail. Female olivish above, lightly streaked brown, with yellower rump and dark brown wings and tail; olive-yellow below. Fairly loud *chip-chip-chip* call, both in flight and when feeding; creaky trilling song. Small, active parties in conifer tops; also descends to ground. Clings sideways and upside-down to extract seeds from cones, aided by unique bill. Found in Himalayan coniferous forests east of Himachal, at 2700–4000m; may descend in winter.

RED-HEADED BULLFINCH *Pyrrhula erythrocephala* 16cm

Tim Inskipp

Male has black face, brick-red crown and nape, grey back, white rump, and orange underparts with black chin and whitish belly and vent. Glossy purple-black wings and forked tail. Female similar, but olive-yellow on crown, with grey-brown back and underbody. Single or double *pheu* call. In small parties, occasionally with other birds. A bird of cover, rather quiet and secretive; feeds in low bushes, sometimes on ground, on seeds, buds, berries, also nectar. Found in forests and bushes in the Himalayas, from Kashmir to extreme east, breeding at 2400–4000m; descends to about 1200m in winter.

BLACK-AND-YELLOW GROSBEAK *Mycerobas icterioides* 23cm

Male has black head, throat, wings, tail and thighs, and yellow collar, back and underbody, with a thick finch bill. Female is grey above, with a buffy rump and belly. Vocal, giving loud two- or three-note whistle. Small parties spend much time on higher branches in tall coniferous forests, hence difficult to see. Feeds on conifer seeds and shoots, also on the ground on berries and insects. Found in mountain forests along the Himalayas, from extreme west to central Nepal, between 1500m and 3500m; may descend in winter.

WHITE-WINGED GROSBEAK *Mycerobas carnipes* 22cm

Male is black above and on throat and breast, with olive-yellow rump, belly and wing-spots and larger white wing patch. Female is brownish-grey where male is black; pale-streaked face and throat. Loud three- or four-note calls, usually given from treetops; occasional harsh notes. Small flocks, often with other grosbeaks; active and noisy. Feeds mostly on juniper seeds and fruit, also insects, especially when breeding. Found in dwarf juniper forest across the Himalayas, at 1500–4000m, but mostly above 2500m even in winter, when may be seen in bamboo and pine.

GOLD-NAPED FINCH *Pyrrhoplectes epauletta* 15cm

Tim Inskipp

Male is blackish, with distinct orange-yellow crown to nape, and white on tertials forming conspicuous V-shape on rear upperparts. Female rufous-brown above, with greenish crown, grey mantle, and grey-black wingtips and tail, and dull brown below; tertials as on male. Has high-pitched *pew pew* call. Small parties of up to ten birds; associates with other finches. Feeds mostly on ground and low bushes, on seeds and berries. Occurs in rhododendron and ringal bamboo forests, from Himachal to the easternmost Himalayas, at 2800–4000m; descends to about 1400m in winter, in dense bushes and scrub.

ROCK BUNTING *Emberiza cia* 16cm

A handsome little bird, having a bluish-grey head with whitish cheeks and supercilium, black lateral crown-stripe, and black eye-stripe and malar stripe meeting behind ear-coverts. Back black-streaked chestnut-brown, fading to rufous-chestnut rump; tail longish, dark brown with white sides; throat and breast blue-grey, rufous-chestnut below. Female duller. Call a repeated *tsi* or *swip*, song a light *tik-chep-tee-zu* with variations; when flushed, flies off straight with *switt-switt* alarm. Feeds mainly on ground, hopping about; eats grain, seeds and insects. Resident in western Himalayas east to central Nepal, to 4000m, on steep rocky hillsides with grass and bushes. Winter visitor to foothills from Tibet and east Nepal eastwards, at 2100–3400m.

WHITE-CAPPED BUNTING *Emberiza stewarti* 15cm

Tim Loseby

Male has distinctive head pattern: whitish (crown greyer), with black eye-stripe, chin and upper throat. Chestnut above, with white outer tail; breast white, with chestnut band below and chestnut flanks, rest of underparts dull fulvous. Female lacks male's head pattern: streaky brown above, with rufous-chestnut rump; buffy below, with rufous breast. Has faint but sharp *tsit* or *chit* call. Small flocks often seen with other buntings and finches. Feeds on ground, chiefly on seeds; rests in bushes and trees. Found on grass-covered rocky hillsides and in scrub, from westernmost Himalayas east to Garhwal, at 1500–3500m; winters in the west Himalayan foothills.

YELLOW-BREASTED BUNTING *Emberiza aureola* 15cm

Göran Ekström

Male has black forehead, face, chin and throat, deep chestnut crown and upperparts, and yellow neck sides and underparts with chestnut breast band and dark-streaked flanks. Black mask absent in winter, when dark ear-coverts and yellow supercilium distinctive. Female dark-streaked brown above, with pale crown-stripe and supercilium bordered by dark stripes, and pale chestnut rump; yellow below, streaked on flanks. Call *tzip tzip*, also a soft *trrssit*. Flocks of up to 40 birds, with other buntings and finches. Feeds on ground, on seeds and grain, or catches insects. Quite common winter visitor from Nepal eastwards, to 1500m, in cultivation and scrub; enters gardens.

GLOSSARY

arboreal Living in trees

axillaries Underwing feathers at the base of the wing, forming the so-called 'armpits'

wing-coverts Feathers on the upper and lower surfaces of the wing that assist streamlining in flight.

crepuscular Active at dusk

deciduous Trees that shed leaves

diurnal Active during daytime

endemic Indigenous and restricted in distribution

evergreen Trees that retain their leaves

eye-stripe Tract of distinctive plumage colour 'running' through the eye

flight feathers Feathers used for flight, mainly the primaries and the secondaries of the wings

fulvous Brownish-yellow

irruption Mass movement of a population from one area into another, usually in response to exhaustion of the food supply

juvenile A young bird in its first full plumage

lores Area immediately in front of eye

mantle The upper part of the back, next to the hindneck

malar stripe Stripe on side of throat

migration The seasonal movement by some species from one area to another, the two regions being well defined and the birds' occurrence predictable

moult The shedding and replacement of worn and damaged feathers at regular intervals

nape Back of neck

nocturnal Active at night

passage migrant A migratory bird that is seen when it stops off to rest and feed during migration between its breeding grounds and its wintering quarters

passerine A member of the largest order of birds, also known as the 'perching birds' because of this ability

pectoral The area of the breast

primaries Main flight feathers

plumes Long, showy feathers often acquired at the start of the breeding season and used for display

race Populations of the same species in different geographical regions showing recognizable differences in plumage and behaviour, but able to interbreed

raptor A term applied to diurnal birds of prey

rectrices The tail feathers (singular: rectrix)

remiges The primaries and secondaries combined (singular: remex)

resident Present within an area throughout the year

rump Area between lower back and base of tail

species Groups of birds that are reproductively isolated from other such groups

speculum Shiny, colourful patch on the secondary feathers of the wings of many ducks

subspecies *See race*

summer plumage The plumage acquired at the start of the breeding season

supercilium The tract of feathers that runs above the eye and eye-stripe as a discrete or distinct stripe

terrestrial Ground-living

wader A category of birds often with long legs and bill, including sandpipers, plovers and curlews

wingspan The length from wingtip to wingtip when wings fully extended in flight

winter plumage The plumage seen during the non-breeding winter months

FURTHER READING

Ali, S. *Indian Hill Birds*. Oxford Univ. Press, Bombay, 1949

Ali, S. *The Birds of Sikkim*. Oxford Univ. Press, Delhi, 1962

Ali, S. *Field Guide to the Birds of the Eastern Himalayas*. Oxford Univ. Press, Bombay, 1977

Ali, S., et al. *A Pictorial Guide to the Birds of the Indian Subcontinent*. Oxford Univ. Press/BNHS, Bombay, 1983

Ali, S. *The Book of Indian Birds*. Oxford Univ. Press/Bombay Natural History Society (BNHS), 12th edition, 1996

Ali, S., B. Biswas and S.D. Ripley. *Birds of Bhutan*. ZSI, Calcutta, 1996

Ali, S., and S.D. Ripley. *The Handbook of the Birds of India and Pakistan, together with those of Bangladesh, Nepal, Bhutan and Sri Lanka*. 10 vols. Oxford Univ. Press, Bombay, 2nd edition, 1978–95

Bates, R.S.P., and E.H.N. Lowther. *Breeding Birds of Kashmir*. Oxford Univ. Press, Bombay, 1952

Das, I. *A Photographic Guide to the Snakes and Other Reptiles of India and Nepal*. New Holland, London, 2010

Das, I. *A Field Guide to the Reptiles of South-East Asia*. New Holland, London, 2008

Dewar, D. *Birds of the Indian Hills*. The Bodley Head, London, 1915

Dewar, D. *Himalayan and Kashmiri Birds*. J. Lane, London, 1923

Francis, C. *A Field Guide to the Mammals of South-East Asia*. New Holland, London, 2008

Grewal, B. *A Photographic Guide to the Birds of India and Nepal*. New Holland, London, 2008

Grimmett, R, C. Inskipp & T Inskipp. *Birds of the Indian Subcontinent*. Christopher Helm, London, 1998

Inskipp, T., et al. *An Annotated Checklist of the Birds of the Oriental Region*. OBC, Sandy, UK, 1996

Inskipp, C. & T. *An Introduction to Birdwatching in Bhutan*. WWF, Thimpu, Bhutan, 1995

Inskipp, C. & T. *A Guide to the Birds of Nepal*. Christopher Helm, London, 2nd edition, 1992

Robson, C. A *Field Guide to the Birds of South-East Asia*. New Holland, London, 2011

USEFUL ADDRESSES

Bombay Natural History Society
Hornbill House, Opp.Lion Gate
Shaheed Bhagat Singh Road
Mumbai 400 001, India
www.bnhs.org

World Wide Fund for Nature India
172B, Lodi Estate
New Delhi 110003, India
www.wwfindia.org

Bird Conservation Nepal
P.O. Box 12465
Kathmandu, Nepal
www.birdlifenepal.org

Bird Link
101/4 Kaushalya Park
Hauz Khas
New Delhi 110016, India

The Oriental Bird Club
P.O Box 324
Bedford MK42 0WG, UK
www.orientalbirdclub.org

INDEX